WATER SHARING CONFLICTS AND THE THREAT TO INTERNATIONAL PEACE

HEARING

BEFORE THE

SUBCOMMITTEE ON EUROPE, EURASIA, AND EMERGING THREATS

OF THE

COMMITTEE ON FOREIGN AFFAIRS HOUSE OF REPRESENTATIVES

ONE HUNDRED THIRTEENTH CONGRESS

SECOND SESSION

NOVEMBER 18, 2014

Serial No. 113–231

Printed for the use of the Committee on Foreign Affairs

Available via the World Wide Web: http://www.foreignaffairs.house.gov/ or http://www.gpo.gov/fdsys/

U.S. GOVERNMENT PRINTING OFFICE

91–455PDF   WASHINGTON : 2014

For sale by the Superintendent of Documents, U.S. Government Printing Office
Internet: bookstore.gpo.gov   Phone: toll free (866) 512–1800; DC area (202) 512–1800
Fax: (202) 512–2104   Mail: Stop IDCC, Washington, DC 20402–0001

COMMITTEE ON FOREIGN AFFAIRS

EDWARD R. ROYCE, California, *Chairman*

CHRISTOPHER H. SMITH, New Jersey
ILEANA ROS-LEHTINEN, Florida
DANA ROHRABACHER, California
STEVE CHABOT, Ohio
JOE WILSON, South Carolina
MICHAEL T. McCAUL, Texas
TED POE, Texas
MATT SALMON, Arizona
TOM MARINO, Pennsylvania
JEFF DUNCAN, South Carolina
ADAM KINZINGER, Illinois
MO BROOKS, Alabama
TOM COTTON, Arkansas
PAUL COOK, California
GEORGE HOLDING, North Carolina
RANDY K. WEBER SR., Texas
SCOTT PERRY, Pennsylvania
STEVE STOCKMAN, Texas
RON DeSANTIS, Florida
DOUG COLLINS, Georgia
MARK MEADOWS, North Carolina
TED S. YOHO, Florida
SEAN DUFFY, Wisconsin
CURT CLAWSON, Florida

ELIOT L. ENGEL, New York
ENI F.H. FALEOMAVAEGA, American Samoa
BRAD SHERMAN, California
GREGORY W. MEEKS, New York
ALBIO SIRES, New Jersey
GERALD E. CONNOLLY, Virginia
THEODORE E. DEUTCH, Florida
BRIAN HIGGINS, New York
KAREN BASS, California
WILLIAM KEATING, Massachusetts
DAVID CICILLINE, Rhode Island
ALAN GRAYSON, Florida
JUAN VARGAS, California
BRADLEY S. SCHNEIDER, Illinois
JOSEPH P. KENNEDY III, Massachusetts
AMI BERA, California
ALAN S. LOWENTHAL, California
GRACE MENG, New York
LOIS FRANKEL, Florida
TULSI GABBARD, Hawaii
JOAQUIN CASTRO, Texas

AMY PORTER, *Chief of Staff*   THOMAS SHEEHY, *Staff Director*
JASON STEINBAUM, *Democratic Staff Director*

———————

SUBCOMMITTEE ON EUROPE, EURASIA, AND EMERGING THREATS

DANA ROHRABACHER, California, *Chairman*

TED POE, Texas
TOM MARINO, Pennsylvania
JEFF DUNCAN, South Carolina
PAUL COOK, California
GEORGE HOLDING, North Carolina
STEVE STOCKMAN, Texas

WILLIAM KEATING, Massachusetts
GREGORY W. MEEKS, New York
ALBIO SIRES, New Jersey
BRIAN HIGGINS, New York
ALAN S. LOWENTHAL, California

CONTENTS

WATER SHARING CONFLICTS AND THE THREAT TO INTERNATIONAL PEACE

TUESDAY, NOVEMBER 18, 2014

House of Representatives,
Subcommittee on Europe, Eurasia, and Emerging Threats,
Committee on Foreign Affairs,
Washington, DC.

The subcommittee met, pursuant to notice, at 2 o'clock p.m., in room 2255 Rayburn House Office Building, Hon. Dana Rohrabacher (chairman of the subcommittee) presiding.

Mr. ROHRABACHER. Subcommittee is called to order. This afternoon's hearing is on the topic that I consider to be of great importance—the sharing and management of water across international borders. When well done, both the environment and the people involved can prosper.

When done poorly or not at all, it can be a cause of conflict. As a proud and longtime resident of southern California, I know first hand the vital importance of having access to clean water.

We drink water. We use water. We use it to generate electricity . Some of us surf in the water. Water is required in a wide range of industries that are essential to the well being of the people, especially agriculture.

Water is a common staple of life, but where it is scarce it is a strategic resource that nations compete to control.

We have before us today a panel of experts who will review the potential for water conflict in two areas of the world—the Aral Sea watershed in Central Asia and along the Nile River in East Africa.

During the Soviet period, water sharing between the five Central Asian republics was commanded by Moscow. The downstream nations of Turkmenistan, Uzbekistan and Kazakhstan needed vast amounts of water, primarily for agriculture.

The upstream republics, Kyrgyzstan and Tajikistan, have a surplus of water and would release their water from their dams during the summer months to irrigate the crops downstream. In exchange, the downstream nations would then send oil and gas and other resources which they possess to the upstream nations during the winter time for fuel and heating.

Since 1991, the Soviet era arrangement between the five republics has broken down. The upstream nations have sought to expand their hydroelectric infrastructure through the building of new dams.

This has been a great source of consternation among the other Central Asian governments, especially Uzbekistan. The breakdown in trust and coordination between these governments concerning water sharing has been a distraction from more pressing issues such as economic development and thwarting radical Islam.

What should be a source of regional cooperation and strength is now a source of regional tension. Let me just note beside the rise in regional strife and waste and misuse of—and overuse of water that has had a dramatic impact—that it has had a dramatic impact on the environment there.

The Aral Sea, which once was one of the largest bodies of fresh water in the world, has all but disappeared. That is a visible tragedy that should be an incentive for the governments of Central Asia to tackle this problem.

In a different part of the world, the Nile and its tributaries are another example of where international cooperation is under stress. The waters that combine to form the Nile flow through ten different countries. It is one of the great rivers of the world and supplies 85 percent of Egypt's water.

Ethiopia is currently executing a plan to construct the Grand Ethiopian Renaissance Dam across the Blue Nile. The Ethiopian Government hopes to have the dam completed by 2017.

They hope that with that accomplishment Ethiopia will become the largest energy exporting state in East Africa. Yet, Egypt has justifiable fears that the new dam will reduce the flow of water that it receives.

The task of finding a solution is complex. Legal agreements governing the control of the Nile's waters date back to the colonial era when many of the current governments didn't even exist.

More recently, multinational—multilateral projects, that is, such as the Nile Basin Initiative, has had some success. But there has yet to be a breakthrough on the largest controversy surrounding the peaceful sharing of the water of the Nile.

In 2013, the rhetoric between Egypt and Ethiopia hit a new low point when then President Morsi stated in not so—it was not so veiled threat, that is, that, and I quote, ''All options are available.'' To respond to the building of the Renaissance Dam it was all of Egypt's options are available and that does sound like a veiled threat.

Since the election, the current government in Cairo, a more positive—there has been a more positive tone and I hope the ongoing negotiations will, at some point, lead to an understanding between these two countries.

I am going to be asking our witnesses to update the subcommittee on the current status of negotiations between Egypt and Ethiopia and lay out what steps our country might take to promote international water cooperation.

The national—excuse me, the natural resources of our planet, including water, are gifts that we can use to improve the lives of all the people of the world. But it is a scare resource and dividing scarce resources is never easy.

But when nations come into conflict over such resources what you also end up with is that you have a waste of those resources, a waste of energy and perhaps conflict that could cause—have a great cost for both sides in such a controversy.

So we hope today to get a little better understanding of these potential conflicts, how they might be averted, so the solutions and we thank our witnesses for coming. And now Ranking Member Mr. Keating for your opening statement.

Mr. KEATING. Well, thank you, Mr. Chairman, and thank you for holding this timely hearing, and I also want to thank you for your leadership on this. I think this committee continues to highlight one of the most important global issues that we are facing and I think that, hopefully, the spotlight we are able to cast on this will result in more interest and more emphasis on dealing with this because, Mr. Chairman, it is indeed one of the most important issues we have to contend with.

So I would also like to thank our witnesses, in particular, Kathleen Kuehnast. Kathleen testified before the full committee on the role of women in conflict prevention and I welcome her back to the committee today. I appreciate your being here as I do all of our witnesses.

Today's hearing topic provides us with an opportunity to look beyond Europe and Eurasia and examine the global impact of depleting resources, climate change, an expanding world population and accompanying social unrest.

Last year, for the first time the Director of National Intelligence, James Clapper, listed ''competition and scarcity involving natural resources'' is a national security threat on par with global terrorism, cyber warfare and nuclear proliferation.

He also noted that terrorists, militants and international crime groups are certain to use declining local food security to gain legitimacy and undermine government authority, and that is what we have to look at if this goes unchanged in the future.

I would add that the prospect of scarcities of vital natural resources including energy, water, land, food and rare earth elements in itself would guarantee geopolitical friction.

Now add lone wolves and extremists who exploit these scenarios into the mix and the domestic relevance of today's conversation becomes clear. Further, it is no secret that threats are more interconnected today than they were, let us say, 15 years ago.

Events which at first seem local and irrelevant have the potential to set of transnational disruptions and affect U.S. national interests. At the same rate, issues of mutual concern provide the opportunity for greater cooperation, and projects that encourage community building and environmental awareness at the local level are taking place and should be encouraged.

In particular, I believe that the women in communities threatened by water scarcity will have an important role to play in the future and should be engaged by their local governments and international communities now.

I agree with Mr. Clapper that the depletion of resources stemming from many factors which, above all, include climate change has the potential to raise a host of issues for U.S. businesses, officials and individuals abroad as well as here at home.

For this reason, Mr. Chairman, I have long advocated for alternative energy sources. Yet, as the representative of what will be, hopefully, 1 day the nation's first offshore wind farm, I deal daily with obstructive businesses and individuals trying to get in the way of this project and others like it in exchange for increasing their companies' profit margins.

I would like to add that, given our distinguished panel of witnesses today and our subcommittee's jurisdiction, I am sure we will be hearing about the tremendous energy reserves in Central Asia and the need for diversifying energy markets.

In this regard, I would like to note that I have and will continue to advocate for the importance of increasing democratic governance and rule of law in that region.

Energy production can get you only so far. I would like to hear from our witnesses on how the United States can engage with Central Asian governments to improve governance, transparency in the energy sector both bilaterally and through international organizations such as the Extractive Industries Transparency Initiative.

However, as we discuss these important issues, I hope that we can continue to keep our own country's movement toward an energy-independent future and the obstacles on this path in our minds.

Thank you, Mr. Chairman.

Mr. ROHRABACHER. Well, I knew you would get global warming in there somewhere. It is okay.

Mr. Meeks, would you like to share an opening statement with us?

Mr. MEEKS. Thank you, Mr. Chairman.

Inequitable access to fresh water is highly variable between and within countries and much of the world's population lives in places where demand for water exceeds supply.

International water shortages are a critical issue with far fetching—far reaching effects on global security and stability. That is why, Mr. Chairman, I am thankful to you and the ranking member for convening such a timely hearing today to address this very, very important topic.

An increasing prevalence of water shortages and the subsequent threats to peace these shortages present are fundamentally a global trend. Indeed, even here in the United States of America we are not immune from this problem.

As some of my colleagues, and you may very well know there was just a program on 60 Minutes, ''Out West,'' talking about how the groundwater is being depleted even in the United States.

So America has a stake in this issue just like every other country and we must realize that all of our futures are dependent upon the actions that we take now.

We have to look at this problem from a geopolitical perspective and understand the complex relations which exist between nations and places like East Africa and Central Asia.

Many of their disputes over water go back decades, if not longer. We must also look at this from a human perspective because at the end of the day it is ultimately about human life. Every year, 2.1 million people, mainly children, die due to illness related to dirty water, poor sanitation and poor hygiene.

One-third of the world's population lives in water-stressed countries, primarily in Asia and Africa. The actions we take today hold the potential to eliminate human suffering tomorrow and promote peace and cooperation for generations to come.

We must think long and hard about how we avoid conflict. We must discuss ways we can advance science and put it to work in the service of mankind.

And most importantly, we must collectively discuss how we can use diplomacy both here in the United States as well as abroad to promote fair, reasonable, responsible and sustainable strategies for cooperation among our friends and partners around the world.

For, indeed, this place that we call Earth is small and we share it with each other and we need to preserve it for each other. Otherwise, we are all subject to perish. So I am grateful, again, to the chairman and the ranking member and I am grateful to our witnesses for being here today and I look forward to hearing the testimony of our witnesses on this very, very important matter.

Thank you.

Mr. ROHRABACHER. Thank you, Mr. Meeks, and I will give a brief introduction to our witnesses, and our first witness is Dr. Paul Sullivan and he is a professor of economics at the National Defense University. For 6 years Dr. Sullivan taught classes at the American University in Cairo.

He obtained his Ph.D. from Yale University, has advised senior U.S. officials on many issues related to energy, water, food, economics, politics and political security issues.

Next, we have Amanda Wooden—Dr. Amanda Wooden—who is an associate professor and director of environmental studies at Bucknell University.

She earned her Ph.D. in international relations and public policy at Claremont Graduate University in California, nearby my home turf. Dr. Wooden served with the organization for security and cooperation in Europe as an economic and environmental officer in Kyrgyzstan.

And finally, we have Dr. Kathleen Kuehnast, who is the director of the Center for Gender and Peacebuilding at the United States Institute of Peace. She is an expert on Central Asia and Central Asian countries like Kyrgyzstan in particular where, for over a decade, she worked as a social scientist for the World Bank with a focus, of course, on Central Asian conflict prevention.

So we are very, very blessed to have you with us today. We thank you for taking your time and sharing your knowledge and experience with us. I would ask if you could try to keep it to a 5-minute summary.

Everything else that you would like to have in a more developed way you could put into the record as part of your—as part of your testimony. And with that, Dr. Sullivan, you may proceed.

STATEMENT OF PAUL SULLIVAN, PH.D., PROFESSOR OF ECONOMICS, NATIONAL DEFENSE UNIVERSITY

Mr. SULLIVAN. Good afternoon, Chairman Rohrabacher, Ranking Member William Keating, from my home state of Massachusetts, the honorable majority and minority members of the subcommittee, Mr. Meeks, it is an honor and a privilege to be giving testimony on this extremely interesting issue.

I heard national security mentioned. This is also involved with economics, politics, geopolitics, diplomacy, possibly the military and also human security, and without the human security of water, food and energy security, terrorism and other things can result.

Before I give any public presentation, I need to give the usual caveats. These are my opinions alone and do not represent the university—the National Defense University—the Department of Defense, the U.S. Government, Georgetown or any other institution I might be associated with.

I will focus mostly on the Blue Nile with the Great Ethiopian Renaissance Dam, or the GERD, in Ethiopia with its potential effects on downstream countries, especially Egypt and the Omo River, with the Gibe Dam cascade and especially Gibe III Dam and the effects on Lake Turkana in Kenya and also in Ethiopia.

These dam systems are way along the way, even as these tensions build. When water is at threat, then energy, food, industry and more are also at threat, including peace and stability.

Severe water shortages can breed poverty, hopelessness, terrorism and even revolution as we saw in Syria with the horrible droughts in 2008 through 2010, essentially sparking off the revolution.

The GERD Dam is built on the Blue Nile, or Abbay, as it is called in the region. Egypt uses about 98 percent of its available water. It is already water stressed.

Water shortages in Egypt built up some tensions leading to the revolution and beyond. Egypt has 55 cubic kilometers of the Nile water for its allocation from a 1959 treaty which Ethiopia does not recognize.

They and others except for Sudan have signed on to another treaty which they recognize. Fifty-five cubic kilometers seems like a lot of water. However, for a population close to 90 million it really is not.

They are living on a knife's edge of water security. When the reservoir behind the GERD is filled it could contain 75 cubic kilometers of water, which is much greater than Egypt's Nile-designated amounts.

It is also one half of the entire water in Lake Nasser, which is the only buffer Egypt has if there is a great shortage of water upstream. The Blue Nile gives Egypt about 60 to 70 percent of all of its water depending on seasonal water flows.

It is very important to understand that the real time to develop water are when it rains the most—in this part of the world it is June through September—and that will likely be when the GERD Dam is filled up. But it is also a time, June to September in Egypt, when it is very hot, also great needs for electricity.

In the 1980s, there were famines and droughts in Ethiopia. About a million people died. At the same time, the water heading toward Egypt was cut back, and when it was cut back the electricity production from hydro dams in Egypt was also cut back.

There was a problem with irrigation. There was a problem with food. Egypt and Ethiopia are intimately connected through the Nile. What happens in Ethiopia can really affect what is happening in Egypt.

If the GERD Dam is filled up too quickly and at the wrong time Egypt could go beyond that knife's edge in security.

This is possibly a nightmarish situation for many Egyptians. They are quite concerned. If they were around in the 1980s they knew what happened then. It is not clear how this dam is going to be filled up. It is not clear from any of the studies.

As a matter of fact, no real studies have been done of this. A lot of ideas have been bantered about the way this might work out but the behavior of the Ethiopian Government on the Omo River with regard to the Gibe cascade is a giveaway.

They could care less, it seems, about what happens to Lake Turkana in Kenya or of the tribals along the river—the Omo River. They have essentially tossed them off their land and there is a huge land grab happening right now.

The Ethiopians have claimed that they are not going to be using the GERD reservoir water for irrigation. However, if there is another famine I can guarantee you they will, and Egypt will be in trouble.

Because just letting the water go through for electricity is very different from using that water for irrigation. It takes more and more of it out. There has to be some way to come to some agreement before it gets worse, before populations grow, before economic demands grow, before agriculture grows along the river.

International law, in this case, is quite weak. There is no enforcement mechanism. You have the Helsinki agreements, the Berlin agreements. You have Article 7 of the U.N. Convention, the Nile Basin agreement and so forth. But without an enforcement agreement, we are really going nowhere with this.

Part of what may be needed, and this is where the United States may walk in, is that we tried to help develop with our allies and partners and many others in the world an enforcement mechanism for these treaties and new treaties to take a look at a better way of sharing water.

If we don't do that, the situation will get more tense along the Nile and many other places in the world.

Thank you.

[The prepared statement of Mr. Sullivan follows:]

Written Testimony of Professor Paul Sullivan, National Defense University, Georgetown University and the Federation of American Scientists to the House Committee on Foreign Affairs, Subcommittee on Europe, Eurasia and Emerging Threats Hearing on: Water Sharing Conflict and the Threat to International Peace, 2255 Rayburn House Office Building, November 18, 2014, 1400-1700.

Good afternoon Chairman Dana Rohrabacher, Ranking Member William Keating, and the honorable majority and minority members of this subcommittee. It is an honor and a privilege to be giving testimony on this very important issue.

Before I give any public presentation I need to give the usual caveats: these are my opinions alone and do not represent those of the National Defense University, the Department of Defense, the U.S. Government, Georgetown or any other entity I may be associated with. I come here as a citizen of this great country who has considerable knowledge and experience in issues related to energy, water, and food security as well as the regions of focus of this testimony.

I will try to keep this at the strategic level at 35,000 feet. Getting into the weeds could redirect this down into discussions that may be less than fruitful. I will focus mostly on the Blue Nile with the Great Ethiopian Renaissance Dam (GERD) in Ethiopia, with its potential effects on downstream countries, especially Egypt, and the Omo River with the Gibe dam cascade, and especially the Gibe III dam, and its effects on Lake Turkana that is partly in Ethiopia, but mostly in Kenya, and the Gibe dam cascade effects on the lower Omo Basin in Ethiopia. Both of these dam systems are well along towards completion even as tensions and other problems loom about them.

This hearing has "water sharing conflict" in its title. However, whenever water security is involved energy security and food security are not far behind. Any proper policy development regarding water needs to be seen within the energy-water-food nexus. The energy-water-food nexus is intimately connected with economic security, human security, national security of a country, and indeed international security – and potential threats to international peace.

Energy extraction, production, processing and even use often requires considerable amounts of water. As an example, the largest use of water overall in this country is in cooling towers of electricity plants. The production of biofuels use massive amounts of water to produce. The production of gasoline and diesel use less water, but still considerable amounts of water. Large hydropower dams require the trapping of massive amounts of water, often measured in tens of kilometers cubed or trillions of gallons, to allow them to produce a certain consistent amount of electricity. Smaller hydropower dams require less distortions to a river environment and will need less water filled behind them in a reservoir. Micro-hydropower needs little or no reservoir development and can easily take care of the electricity needs of small village or towns that dot, for example, Ethiopia, especially given that Ethiopia is a mostly rural and fairly spread-out population.

Geothermal power diverts much less water per kilowatt hour than large hydropower systems. Ethiopia has significant geothermal energy resources in many parts of the country given that it has many volcanoes and is part of the Rift Valley system, a geologically hot zone. Geothermal water use also will not divert large river systems causing potentially massive tensions with downstream populations and countries.

Ethiopia has considerable solar and wind energy potential and may have not insignificant natural gas reserves, although this latter resource's potential importance is yet to be settled. Ethiopia also has the potential to use energy systems that take advantage of temperature gradients along the sides of mountains and plateaus, given its fascinating and varied topology.

Ethiopia has massive hydropower potential and is using only about 3 percent of it. Only about 20 percent of its people has access to electricity. Ethiopia is not only poor, it is energy poor. Ethiopia is looking to move its country forward via energy and other developments, but there are other ways to develop energy sources than those that mortgaging large amounts of future GDP to pay off domestic and international debt for immensely expensive hydropower project. The GERD could end up costing well over \$5 billion. The numerous dam projects, including the Gibe cascade will end up costing many times the GERD. Ethiopia's GDP is about \$50 billion at best and its GDP per capita may be about \$500. Yes, this is a poor country pouring billions into expensive infrastructure financed with either financial or political debts to financing states, such as China.

Instead of building massive, and seemingly oversized hydropower dams at a rapid rate, and in the shadows of secrecy and surprise, it may make more sense for Ethiopia to focus more on its other significant sources of energy and be more open about its plans with its neighbors and others. If there is one thing that makes me wonder almost as much as the potential for severe water stress in the downstream countries in the future it is the way that Ethiopia is handling the public relations and research on its massive dams, such as the Great Ethiopian Renaissance Dam.

The reports on this dam are often vague at best on the very important issues of when the reservoir will be filled and how it will be filled, amongst other issues. Environmental and social and economic impact reports are either lacking or insufficient. The sorts of reports private investors and potential funders, such as The World Bank, would require are not sufficiently there. Many international financial institutions backed off from supporting the GERD, or were never asked to support the GERD. The Ethiopian government claims that they and the sale of dam bonds, along with some aid from China, will make this happen. China does not ask for the sorts of environmental and social impact statements that The World Bank would require. An Italian company, the Chinese, and others to be named, it seems, will be developing the electricity infrastructure around the dam to connect it to the rest of the country and to, possibly, Kenya, Sudan and there is talk of connecting even to Egypt, which seems to be a bit of stretch at the moment.

The initial building of the dam started in the midst of the first revolution in Egypt in 2011, when Egypt was at its political weakest and was focused almost entirely on its internal conflicts and tensions. The timing of this dam is one of the many reasons why Egypt in particular is upset about it. Many Egyptians perceive that the Ethiopians started building this dam when they were down on the ground during one of the most difficult moments of its recent past in order to spite them, and to get around their historical rights to the Nile water. Water, food and energy shortages were some of the sparks of the rebellion. Resource security is a very big deal for Egypt and Egyptians.

Ethiopia is a poor country. It had famines in the past that shocked the world. About one million Ethiopians died in the last famine. Its political history is checkered with the sort of top-down, brute-force approach that has been further exemplified by the GERD and on the Gibe dam cascades on the Omo River. There seems to be little pushback from the communities in Ethiopia that may be harmed by these projects. There seems to be little freedom of speech on the issues related to the dam inside of the country. The upcoming elections in 2015 could be a bit more contentious than otherwise due to the vast infrastructure programs in the country -- the hydropower dams being a part of that.

Ethiopia has been growing at about a 10% rate in its GDP in recent years. Much of this growth has been from government expenditures in infrastructure. Much of the funding for this infrastructure has been from debt. One wonders where the Ethiopian economy may be headed in the next few years with this growing burden of especially domestic debt. Bond issues to Ethiopians are debt issues to Ethiopians. The bond rates are from about 1.7 percent to about 2.5 percent depending on whether they are in Euros of US Dollars and the maturity of the bond. This may seem low, but where is the income to pay these back to come from. Also, I expect the future bond needs to have higher interest rates. And the dam systems to be built will need more bonds. The GERD may also need more bond issues.

Ethiopia is betting its future on debt-financed massive hydropower projects that will likely cause even greater tensions in the future as climate change or, if you prefer, weather pattern changes, kick in. The capacity of the dams, their electricity production, and the potential for irrigation will be determined by the rains of the future. According to many those rains may be less certain then they have been in the past. Climate change and weather pattern changes could have deep impacts not only on Ethiopia, but also on its downstream neighbors of its many rivers Sudan and Egypt to the North, Kenya to the South, and Eritrea, Somaliland and Somalia to the East and southeast.

Ethiopia is sometimes called "The Water Tower of Africa", but there is just so much water to go around and it may be less in the future. Yet, the future will also likely bring population, industrial, agricultural and other growth to those along the rivers and those who rely on the groundwater and rains that develop from the river waters out of Ethiopia.

Ethiopia needs water for its population growth and it has a larger population than Egypt. Ethiopia's population is also growing faster than Egypt. Ethiopia also needs water for increased industrialization, which it plans, and more irrigation for its crops, as well as more water to produce energy. Ethiopia is in a serious drive to build hydropower dams. The most important one for the flows of the Blue Nile to Egypt is the Great Ethiopian Renaissance Dam or GERD. As I said, Ethiopia will need the energy from this dam, which is name plate of about 6000MW, but actual expected capacity of much less. Ethiopia may also need the irrigation potential that will come out of filling up the reservoir behind it.

Ethiopia claims that no irrigation will occur around the GERD and its reservoir. One of the arguments is that it is too mountainous near to it. However, Ethiopia has a history of droughts, near droughts and famine. I can see a scenario when the water from this massive dam could be redirected toward agricultural and other needs. Politically it would not be possible for Ethiopia to not use this water in times of great need, and Ethiopia has a past of unpredictable rains at times – with massive death tolls as a result.

Ethiopia as set up some very large agricultural areas with a lot of excellent land set aside for foreign companies to grow food, cotton, biofuels, etc. Ethiopia has been one of the main sources of ""land grabs" by companies and countries from Asia, the GCC, and more. Sugar plantations and other very large agricultural developments are occurring along the lower Omo River in Ethiopia.

The Omo River is the main source of water to replenish Lake Turkana in Kenya. The Ethiopian government's behavior with regard to Gibe dam cascade of five large hydropower projects, and especially the filling of the Gibe III dam, could be seen as a warning for what may happen on the Blue Nile or Abbay, as it is known in region, where the GERD project can be found. The seeming complete disregard of the poorest of the poor in the Omo River basin, and on those on shores of Lake Turkana, and those who rely on Lake Turkana for their livelihoods and existence, is an object lesson for all involved. Kenyans in this area could pay a very high price for these irrigation and energy projects in Ethiopia. Also, the electricity production from the large hydropower projects in Ethiopia in the northwest, such as the GERD and the southwest, such as Gibe, are unlikely to help the local populations much. Most of this electricity will be sent by high-voltage transmission lines to the larger cities and outside of the country it seems. Ethiopia is a mostly rural country. This may change rapidly with the rapid and intense electrification of the country. All of the positive and negative aspects of quick urbanization will result. I do not see much planning being done for balanced growth. There seems to be a headlong and headstrong drive towards "modernization at all costs". Ethiopia may find itself in an unstable social and political environment internally unless they rebalance the way they are looking at economic development. Electrification and infrastructure development is normally good for a country and its people, but this needs to be done properly and in tune with other aspects of the economy and society.

Water is a growing problem in Egypt. The Nile has historically declining flows of water. The population of Egypt is on the rise. Egypt will need more water for that growing population and, hopefully, its growing industry, agriculture and service sectors. Getting more people employed will likely mean more water needs.

Egypt can do a lot to increase the efficiency of its use of water in industry, agriculture and in the households, and also in energy extraction and production, but it can go just so far. Those efficiency improvements will also take time.

The Nile goes from north to south. Most of Egypt's water comes from outside of the country – about 96 to 98 percent. Most of it, about 65-75 percent of so depending on the season, comes out of the Blue Nile which originates in Ethiopia.

Sudan with a population much smaller than both Ethiopia and Egypt will also need increasing amounts of water for its growing population and other needs. Sudan expects some return from the GERD given the project is just 25 or so kilometers from its border with Ethiopia. It expects electricity, better water control and maybe even some irrigation benefits from that better water control. Again, these benefits are so far publicly unclear.

One of the tensest moments for the GERD project will be during the time that the reservoir will be filled. The slower Ethiopia fills the reservoir the slower will Ethiopia ramp up its electricity and it's so-far publicly denied irrigation potential. The faster Ethiopia fills up the reservoir the more stress will be put on Egypt and Sudan, for not only irrigation and other uses of the Nile water, but also for hydropower electricity production.

Hydropower dams in Sudan and Egypt will be producing electricity at lesser rates than otherwise as the filling happens. Egypt, of course, could use more of the water from Lake Nasser to increase the flows. However, there would depend on the existent heights of the water behind the Aswan High Dam and how much of that water can be used for hydropower. So we can see a direct connection between electricity production in Egypt and Sudan and the filling of the GERD project reservoir in Ethiopia. How the electricity production in these downstream states will be effected depends on when the filling occurs and how quickly it occurs.

On the other hand, Egypt has its Toshka Project that extracts more water from the Nile for irrigation developments in the "New Valley". This is essentially a new Nile developing westerly of the present Nile and going right into the Western Desert to develop those lands and move large groups of Egyptians from along the Nile to the

Western Desert, an otherwise fairly unpopulated place of a somewhat hostile natural environment. When Egypt started this project during the 1990s Ethiopia was quite upset about this.

The history of Egyptian-Ethiopian tensions about Nile waters go back many centuries. There were times of increased tensions when there were any discussions of putting dams in Ethiopia. The 1922 and 1959 Nile waters treaties were signed without Ethiopia even part of the negotiation, and gave all of the water of the Nile to Egypt and Sudan. In the latter one Egypt got 75 percent and Sudan got 25 percent. The rest of the countries of the Nile Basin were not even in the discussions. The Aswan High Dam in Egypt and two dams in Sudan were built based, in part, on the 1956 treaty. Most of the other lands that were part of the Nile Basin did not have their independence until after this treaty was signed.

The building of the Aswan High Dam helped prompt the Orthodox Church of Ethiopia to split from the Orthodox (Coptic) Church of Alexandria in Egypt. There were other reasons, but many think the building of the Aswan High Dam was the last straw. Inter-religious as well as international and other intergroup tensions have been a part of the Nile since there were people living on and near to it.

The tensions due to water issues were often focused on the White Nile. But all along they should have focused on the Blue Nile, where most of the Nile River, which is the confluence of the White and Blue Niles beginning at Khartoum, Sudan. In some parts of the year there are larger or smaller percentage from the Blue Nile given that the heaviest rains that fill the Blue Nile are in the June-September time periods, with the strongest rains in the July and August time periods. There are times of the year when the rains are quite low in Ethiopia and the Blue Nile flows are much less. When the rains a low or there is a drought in Ethiopia, or that water is not flowing so much out of Ethiopia, then Sudan and Egypt see the effects.

In the 1980s when the devastating droughts hit Ethiopia the effects could be felt all the way to the Mediterranean because of the overwhelming importance of the Blue Nile for the total amounts of water in the Nile proper flowing through Sudan and Egypt. Egypt was saved by the water stored in Lake Nasser. However, even with Lake Nasser as the height of the lake went down the flows out of it slowed down. Electricity production slowed way down and agricultural and other problems resulted. What happened during the Ethiopian droughts in the 1980s in Egypt is resonating today because the reservoir behind the GERD could be as much as 65-75 cubic kilometers, which is much greater than the entire Nile flow allocated to Egypt in the 1956 treaty, about 55 cubic kilometers. The volume of Lake Nasser is just about 132 cubic kilometers, or only about twice the maximum fill volume of the GERD. One can see why the Egyptians are wary of how and over what time period this reservoir is to be filled behind the GERD. If the fill cuts too much into the downstream flows too quickly there could be extreme energy, water and food stress for the downstream countries.

Sudan also has plans to develop more hydropower and more irrigation as its population grows and its needs increase. One wonders how they will fit into this series of building international water tensions. Their leader General Al Bashir says that he is ok with the GERD project. One has to wonder what is behind that statement.

Egypt and Ethiopia, as well as Sudan need to come to some accord to figure out where all of this is going. The earlier accords that gave all of the water to the Nile to Egypt and Sudan, and yet left all the other Nile Basin countries out of the treaties can no longer work. The Nile Basin Initiative was started in 1999 to develop greater cooperative use of the Nile waters for all of the states along the Nile. Membership includes Egypt, Ethiopia, Sudan, South Sudan, The Democratic Republic of Congo, Tanzania, Uganda, Burundi, Kenya, and Rwanda. The Nile Basin Initiative seems to not be the place to resolve the Ethiopian-Egypt spat over the Blue Nile and the GERD project. It also does not seem to be the place to resolve the looming issues around the Gibe dam cascades and Lake Turkana.

The Cooperative Framework Agreement (CFA) was signed in 2010 Ethiopia, Tanzania, Uganda, Kenya, and Rwanda, under great objection from Egypt and Sudan. Burundi and joined later. South Sudan has exhibited interest in signing, but is in the midst of violent conflict and this has gotten in the way of its moving ahead on this. One of the main reasons they signed the CFA was to try to even the use of the Nile water and rights to water and the concomitant benefits from that water. The tensions on the Nile are not just from the Blue Nile, its many tributaries, and the GERD project, but also extend into the White Nile and its tributaries, which are far more important to most of the countries other than Egypt, Ethiopia and Sudan given that much of their Nile water comes from the White Nile and its tributaries and sources, as well as much better rainfall patterns than can be found in Sudan and Egypt. The international water tensions of the Nile can be separated to a great extent between those who rely on the White Nile mostly, those who rely on the Blue Nile mostly, and those who rely on both.

The tensions over water, especially in the Sudd, a large swampy area in the present-day South Sudan led to violent conflict between the northern and southern parts of the formerly unified Sudan. John Garang, the former head of the rebel group, the SPLA, wrote on the problems of water in the Sudd as part of his academic training. Many believe the second Sudanese Civil War was sparked by the attempts of the northern part of Sudan to dig the Junglei Canal through the Sudd to bring more fresh water to the north and bypassing or neglecting the southerners.

Water can lead to war in this region. As populations, economies, food and other needs grow in Ethiopia, Sudan and Egypt tensions along the Blue Nile will increase. There is just so much water to go around. The sources of most of the water for the Blue Nile are in Ethiopia, but some of the greatest future needs for the water can be found in both Egypt and Ethiopia. How this all works out could be determined as much by hydrology as it is from the relative economic, diplomatic, military and other clout of Egypt and Ethiopia.

Egypt is now quite a bit stronger than Ethiopia militarily and diplomatically, but that power has been declining in relative terms for some time. Ethiopia is not the weak state of famine-stricken people and hundreds of rebel groups that it once was. It is still a poor country, but it seems to be on the upswing with power both inside and outside of the country.

Egypt is also caught up in its own internal insurgency, a terrorism problem in the Sinai, and increasing threats out of Libya, Gaza and more. It has many more things to focus on than just the GERD, yet it seems to have periodic laser-beam focus on this issue given the importance of water security to the country. Egypt could destroy the GERD dam and many other dams to follow it, and some that were even built before it.

The questions are: (1) will they and (2) would that make any sense? If the filling of the GERD causes serious water, food and energy issues in Egypt then, frankly, it would be too late to do much of anything. Blowing up a dam with 75 cubic kilometers of water behind it would be a disaster in so many ways that it would be unthinkable. Threats from Egypt could be emotionally real, but the practical aspects of such an attack make it unlikely. Given Egypt's perilous financial and economic situation this makes war all the more unlikely, unless they can get a couple of Gulf States to fund it. However, the Gulf States have other worries to deal with, such as Syria, Iraq and their own internal stresses for some of them that are far to the top of their concerns. The Nile does not seem to be high up in their list of concerns -- other than the concerns about the lands they are leasing and using along the Nile River Basin. Egypt has a powerful military, but its military also seems to be thinking strategically and in the longer run. They may prove to be more cautious and cunning on the issues of the Nile River, the GERD and other issues than some may imagine.

Also, Egypt has been making lots of conciliatory moves towards Ethiopia lately with trade agreements and other diplomatic accords. It seems that they may have to agree to the dam, but they may be on the hunt for other concessions on issues of importance to them to make the "accepting" of this difficult issue easier. Again, the GERD started in the midst of the Egyptian Revolution of 2011 when Egypt was too introspective and weak to put a stop to it. By the time it got to focus on the dam it really seems it was too late.

Now the negotiations for concessions and tradeoffs begin. These negotiations could involve issues related to Sudan, South Sudan, Israel, Somalia, Eritrea (Egypt backed them in their war for independence against Ethiopia), Yemen, the GCC, and, possibly, the increasing clout of Ethiopia in Africa. The negotiations will most likely include how quickly and how much the reservoir behind the dam is filled, and at what times it is filled, as well as water flows thereafter and whether any of this water will be used for irrigation or other activities that may further drain off water that could go to Egypt. Egypt is in a tough position on this. However, Egypt has been involved in East African affairs for some time, and could become problematic for Ethiopia in indirect ways if the need arises in the Egyptian perspective – or if it is needed during tougher parts of the negotiations.

There need to be better studies of the downstream and upstream effects of the GERD. More on the downstream than the upstream given the risks of conflict. Egypt needs to rethink its water use, especially in flood irrigation and the types of crops it grows. Water foot prints need to be considered by all parties involved. And some policy changes will likely need to be developed on the use and in the charging of water. There may also be some efforts put into the development of a regional power grid, where all countries will benefit from the inevitable development of hydropower in Ethiopia.

Hydropower dams are also quite dangerous in risky environments. The Geneva Accords have a special section defining the law of war with regard to how to deal with nuclear and hydropower stations. Reservoirs with trillions of gallons of water hold massive potential for damage. This is particularly relevant if water tensions turn to water wars. As we saw fears associated with the tries at the Mosul and Haditha Dams by ISIS.

Water could lead to conflict on the Nile. The most likely conflict is between Egypt and Ethiopia, but if it does happen it would most likely be a proxy conflict not a direct one. Egypt is under water stress. Sudan is under water stress. South Sudan has massive excesses of water as do some others along the Nile Basin. There may be changes there in the future for water trade with potentially huge income streams as the tensions on the other parts of the Nile increase.

Populations, industry, agriculture and more will grow over the years within the entire Nile basin. Either better water sharing and water use policies are developed or there could be real trouble. A lot of this boils down to virtual water trade and direct water trade. It also involves understanding and using water footprints better. Above all it requires learning to get along with shared resources.

Out in the west of the US it is said that whisky is for drinking, but water is for fighting. In this part of Africa and the Middle East and North Africa water can also be for fighting. I am reminded of the scene from the movie "Lawrence of Arabia" when Sherif Ali guns down Lawrence's Bedouin guide and states that the Bedouin was nothing and the water was everything.

With climate change barreling forward and with water needs growing more needs to be done on such issues. If there are to be proper treaties then they need to be set up for the long term --- 15, 20, or even 50 years. They need flexibility to take into consideration unpredictable events and changes. These treaties also need to take into consideration that Ethiopia, Egypt, Sudan and others in the region are on their way, hopefully, to better economic and human development. And this means likely more water demand.

And what will all of the changes coming in the future in weather, the climate and human and economic development mean for the Nile, for the Omo, for Egypt, Kenya, Ethiopia and others who may be involved with the tensions building over water in East and Northeast Africa? This is still uncertain, yet the future tensions and conflicts that could erupt from these changes could make today's tensions and potential conflicts seem tame.

It is clear that the Nile Basin Initiative and the Cooperative Framework Agreement (CFA) are not working in the way they were initially intended, and not all want to be a party to the agreements. The CFA is impotent to resolve the disputes between Egypt and Ethiopia on the GERD project and other dams that are in the plans. Egypt is not a signatory to the CFA and does not recognize it. Ethiopia does not recognize the 1959 treaty that Egypt thinks is valid in these circumstances. Egypt calls forth Article 7 of the UN Convention on Non-Navigational Uses of International Watercourses which describes "the obligation not to cause significant harm", and if harm does occur the country that caused the harm must mitigate that harm. Ethiopia is calling on the principle of equitable shares. It is a good thing that Ethiopia is not calling on the concept of complete sovereignty of internal resources. Now that would be a real mess.

International water law as it stands has no clout to alleviate the problems that will likely arise for various groups in the lower Omo River area and along and near to Lake Turkana in Kenya and Ethiopia. It seems the problems of the GIBE cascade of dams, and especially the GIBE III dam and its related irrigated plantations would fall under Article 7 of the UN Convention on Non-Navigational Uses of International Watercourse, but it also seems that Ethiopia is the more powerful party in the Omo River issues than Kenya. Maybe might really does describe "right" in water disputes. Ethiopia and Kenya are negotiating as the dam cascade system continues to be built and the irrigation schemes are leased and handed out in Ethiopia.

International law on the use of water seems very woolly indeed and is full of all sorts of holes through which many countries can walk. The Helsinki Agreements and Berlin Rules have been batted around the Nile water issues between Egypt and Ethiopia without much clarity and decisiveness. Also, how do states resolve an issues that involves the virtually unfettered flow of water to Egypt from Ethiopia since the time of the Pharos? How do countries looking forward to economic and human development in a limited water environment figure out how to share that water properly, however "properly" might be defined via negotiations, rather than war, hopefully. How could such countries figure out more efficient and effective ways of jointly using the shared and limited water resources? These and many other questions loom over the future of the Nile Basin, the Omo Basin and many other river basins, aquifers, and other sources of water in the region. Some of these basins and aquifers are connected to others in complex ways, but that is for another hearing.

International law on water has no real effective enforcement mechanism. It is based on customary laws of the past and treaties and agreements that read well, but are not enforceable. Maybe the UN Security Council could get involved. However, given the increasing power and alliances of Ethiopia this may go nowhere. It is clearly time for the world community to develop some enforcement mechanism for laws regarding international water courses as we head into times of even greater water stress.

Why is this important for the United States? Consider the region we are discussing. Egypt, Ethiopia and Kenya are important allies and partners for the United States for various important issues such as peace and security in North Africa, the Sahel, the Horn of Africa and, indeed, in the Nile Basin. The U.S. has interests in these areas including, but not limited to: anti-terrorism efforts, the "Peace Process", freedom and security of the sea lanes of communications (SLOCS), health issues, humanitarian issues, energy security and trade security, and so much more. The United States is also focusing a lot more on Africa, a potential economic, political and military giant in the future as it finds its way towards a new future. The United States should be a positive part of that future.

Potential policy options for the U.S.:

Act as a convener, along with other stakeholder, of the parties involved with the water tensions in the region

Tie aid to the treatment of persons, groups, and tribes who will be displaced by hydropower systems.

The US and other powers could work more on developing an international system for understanding and resolving resource conflicts in the energy-water-food nexus

Considering that the Nile has competing treaties then a new set of treaty obligations could be developed that would incorporate all stake holders in the Nile Basin

The U.S. and others could develop an international coalition for the development of efficiency improving technologies for the energy-water-food nexus

The private sector in the U.S. could be encouraged to be a part of the solution to future stresses and tensions in the energy-water-food nexus in this region and others

There could be a more effective global effort to reduce the water footprints of energy and food in particular, but also of other activities and uses for water

The U.S. and others could put more effective efforts into the analysis and mitigation of land grabs in this region and others. This should also apply to water grabs, which are in effect what happens in a land grab.

As the world faces new tensions growing out of reduced water per capita supplies then more efforts technologically, diplomatically and in business could be made to help alleviate the tensions that seem so far inevitable in the future.

Pressure could be put on the parties involved in such water disputes to be a fair and as equitable as possible

The U.S. needs Ethiopia as a partner in issues related to the Horn of Africa and other parts of the region and its connected regions. However, this should not push the U.S. to take pressure off of issues that may cause far bigger problems in the future

New dam designs need to be developed for water stressed areas with greater emphasis on the effects on the downstream users of water, sediment movements, habitats, the treatment of local populations, and seismicity.

New energy diversity portfolios need to be developed for water-stressed areas and for areas that may face increasing water stress in the future

Different cropping and irrigation patterns and technologies need to be developed in the Nile Basin region, especially in the tensest areas.

The U.S. will need to work more closely with China and others who are big lenders, donors, and investors in the region to figure out better long term plans for economic and human development whilst encouraging a better development of the understand and effectiveness of the energy-water-food nexus

As the relative power relations between Ethiopia and Egypt develop the U.S. needs to keep a close eye on not only the direct relations between them, but also the indirect and proxy relations related to these tensions.

The U.S. is a major aid donor to Ethiopia. Ethiopia is an important part of our Powering Africa program. It is also an important part of our outreach to Africa. The U.S. is also a significant donor to Egypt and works with Egypt on many sensitive and important issues related to the national security and other interests of the U.S. We need to tread carefully in the mutual development of our relations with these two important countries that often bridge Africa and the Middle East in many complex ways.

The U.S. could take a closer look at what is happening on the Omo River, how this could affect the people of the Omo River Basin, the people of Lake Turkana, and the very existence of Lake Turkana in Kenya. Kenya is also an important partner of the U.S. on many of our initiatives in the region.

The U.S. could take a close look at the water relations between Ethiopia and Somalia, South Sudan, and Eritrea. These were not part of this hearing, but could be part of a future hearing.

The U.S. could look into how it could connect the energy-water-food nexus and the nexus policies that could result from thinking in nexus ways into how it develops its relations with Egypt, Ethiopia, Sudan, other countries in the Nile River Basin and other countries in Africa. (This could be applied to our activities in other parts of the world as well, where were almost always separate out our policies on energy from those of water and those of food.)

(Sources supplied on request. Any further questions from the committee or others may be directed to Professor Sullivan at DrSullivenergywaterfood@gmail.com)

Mr. ROHRABACHER. Dr. Wooden.

STATEMENT OF AMANDA WOODEN, PH.D., ASSOCIATE PROFESSOR OF ENVIRONMENTAL STUDIES, BUCKNELL UNIVERSITY

Ms. WOODEN. Thank you. Mr. Chairman Rohrabacher, Ranking Member Keating, Member Meeks and other distinguished members of the committee, thank you for inviting me to testify here today.

In Central Asia, the main water disputes concern the water energy nexus but a direct relationship between water scarcity and interstate conflict is an unlikely scenario.

There are already existing contentious water politics within several countries in the region and these combine and most recently are combining with finger pointing between Central Asian leaders who, at times, have used nationalistic rhetoric and threats about water to implement other political issues.

However, existing cooperation, even within the current weak institutional regional water sharing network, means that conflict is avoidable. So to help strengthen cooperation, the U.S. Government should expand support for renewable energy development and electricity distribution, help tackle pernicious pollution problems, continue and expand support for scientific research in and about the region and increase support for climate change mitigation and adaptation.

The most important regional water tensions and social discontent issues are hydro electric dam development, irrigation management, infrastructure failures, militarized border zone water sharing and flooding potential.

Aral Sea ecosystem collapse is a significant livelihood threat. The biggest future of water supply risks are glacier loss and precipitation changes as well as hidden creeping pollution problems and industrialization.

Kyrgyzstan and Tajikistan currently face significant adaptive capacity limitations and all countries need to better tackle rural vulnerabilities and deal with citizens' everyday water challenges.

The glacier-fed rivers Syr Dar'ya and Amu Dar'ya, the main rivers in the region, terminate in the former Aral Sea, which, as the chairman already described, was once the world's fourth largest lake and now is only 10 percent of its former size.

Most recent NASA images show the largest lake portion no longer exists. I gave you an appendix of images, about 15 images, that show you the map of the region and what this transition looks like.

The Aral Sea collapse and continuing decline of residents' quality of life, which includes access to fresh water, soil quality loss, shortened growing season length, could contribute to dissent in Uzbekistan, and this is important to understand.

The ongoing disagreements between Central Asian governments over the Syr Dar'ya and the Amu Dar'ya and seasonal tensions in securitised cross border communities like the Ferghana Valley are the key tension points.

There are also complex functioning relationships from the local to the national level. When we talk about conflict it is important

24

to recognize that cooperation is more regular than disagreements and tense events, although that cooperation seems fragile at times.

So we have moments where paying attention to what can work in cooperation and how we can enhance that is important. The biggest conflict risks in Central Asia are political and economic—government willingness to tackle everyday struggles with water and power, authoritarian state treatments of information sharing and dissent—human rights and access to water being one of those—subsequent contention between people and governments about their nonresponsive policies which we see in times of drought and responses, and regional leaders' use of nationalistic rhetoric to lay claim to waterways and rationalize particular waterway uses are—these are the biggest risks.

I also would suggest caution when we use danger and risk rhetoric. This language can contribute to a difficult—already difficult dialogue between Central Asian nations. So more specifically, let me talk about hydro energy.

Around 90 percent of Kyrgyzstan and Tajikistan's energy production—electricity production is from hydro power so both countries are investing heavily in reviving Soviet-era plans to expand this sector.

Kyrgyzstan is constructing the large Kambar-Ata dams I and II and four smaller dams on the Naryn River with the completion date target of 2019, and Tajikistan is heavily investing in the Rogun Dam, which will displace more than 40,000 people.

The Government of Uzbekistan has seemingly engaged in economic retribution through border securitisation and cutting off gas supplies to both countries at various times and the downstream countries perceive this as a direct response to dam construction.

Southern Kyrgyzstan's gas has been cut off since spring. So both Uzbekistan and Kazakhstan continue to pressure for halting this large-dam construction or engaging in regional evaluation of the projects mutually.

They are concerned these dams will impact availability and timing of flow for downstream irrigation-dependent and flood risk communities.

However, both Kyrgyzstan and Tajikistan face the reality of electricity shortages and impact of power outages on their population's well being and popular discontent. Recent research that I conducted in Kyrgyzstan concerns the 2010 revolution, which was in part sparked by protest about electricity blackouts, partial privatization of this sector, and increase in tariffs. This year, 2014, is a very similar year to 2008 and 2009 in Kyrgyzstan.

Already blackouts are likely for the winter, tariffs are being discussed and it is hard to imagine for Kyrgyzstan and also Tajikistan that they would halt the controversial dam construction projects given their severe domestic electricity shortages.

So how do we think about tackling this? Moving forward, there needs to be clarity about how Rogun, the new Rogun and Naryn dams, will be operated or tensions will remain and perhaps increase during dam construction and the reservoir-filling period.

The CASA-1000 project to which the U.S. Government is contributing to export electricity from Kyrgyzstan and Tajikistan to Afghanistan and Pakistan is another point of contention.

Uzbekistan opposes this project as it arguably depends on these new dams for adequate electricity export. Perhaps one way in which the United States Government could help to improve relations is following the International Crisis Group's suggestion to support creating multiple bilateral agreements instead of the current dysfunctional multilateral agreements. So this has happened between Kazakhstan and China, between Kyrgyzstan and Kazakhstan in the Chui-Talas River Commission, and Kyrgyzstan and Tajikistan could engage in bilateral agreements.

Of course, we would hope that Uzbekistan and Tajikistan and Uzbekistan and Kyrgyzstan would do the same but those are less likely.

A key future change is projected decreases in the Tien Shan and Pamir mountain glacier surface areas and volume over the next few decades. The region will shift from a glacier-dependent hydrological regime to a precipitation-dependent one, resulting in greater variation in water levels, seasonally and annually.

So adapting to this slow-moving process requires significant international support. I have five key policy suggestions.

The first is, given this tight relationship between water and energy and regional distribution disputes as well as household energy and security, it is most important to invest in expansion of renewable energy sources, off grid household energy systems and related infrastructure.

The USAID Energy Links program works on institutional strengthening and some energy efficiency improvement. However, much more is needed.

Second, water pollution is often missed in discussions of conflict but it is fundamental for tackling everyday problems residents face and consequentially for adjusting and maintaining political stability.

For example, U.S. Government funding to help address the uranium tailings legacy would be positively received by multiple countries so it would be cooperation enhancing and something we want to return to.

Large-scale industrial projects should be monitored, especially, for example, I can mention Kumtor gold mine which is operating on four glaciers in the Tien Shen mountains.

There are several glaciers that are actively being mined in the operation of this gold mine facility and this is something that should be evaluated when we talk about the EITI, for example, that Mr. Keating mentioned.

Third, supporting Central Asian glaciology and hydrology research and scientific monitoring is a valuable contribution, and USAID is considering contributing to the World Bank Central Asia hydro meteorology modernization program and I would support this.

And, finally, glacial decline is already happening and it will have clear impacts on regional water distribution. It is necessary to increase funding for mitigation and adaptation, improve and expand programs such as the USAID wheat resiliency program and tackle this issue more broadly.

Thank you very much.

[The prepared statement of Ms. Wooden follows:]

Statement to United States House of Representatives Committee on Foreign Affairs
Subcommittee on Europe, Eurasia, and Emerging Threats hearing, 2:00 p.m., November 18, 2014
"Water Sharing Conflicts and the Threat to International Peace"

Testimony of Amanda E. Wooden, Ph.D. (Associate Professor of Environmental Politics & Policy, Bucknell University)

Chairman Rohrabacher, Ranking Member Keating, other distinguished Members of the Committee, thank you for inviting me to testify here today. I will provide an overview of water security in Central Asia, potential for regional conflict, how water-driven disputes impact U.S. interests, and possible steps for the U.S. government.

Summary – Strategic Overview

In Central Asia, a direct relationship between water scarcity and interstate conflict is an unlikely scenario. The main water disputes concern the water-energy nexus. Existing cooperation, even with the current weak regional water-sharing institutions, means that conflict is avoidable. It is necessary to understand and evaluate the long-term, complex and indirect relationships between water and contentious politics *within countries* in Central Asia. Internal (intrastate) instability may be indirectly related to water regime changes, but it is also a consequence of socio-economic changes we see happening in the region (namely migration and poverty). Internal political problems related to water – such as drought driven hydroelectricity shortages - can spillover into international disputes. The already existing contentious politics of water *within* several countries in the region combines and is enhanced with finger pointing between Central Asian leaders, who at times have used threats about water as political instruments to influence other issues.

Key Issue Overview

In this overview of regional water relationships in Central Asia, I consider and focus on the five 'post-Soviet' countries in the Aral Sea basin: Kazakhstan, Kyrgyzstan, Tajikistan, Turkmenistan, and Uzbekistan (see map 1). Shared water issues in these countries are connected to China and Afghanistan, who are on the edges of the Aral basin (see map 2).

Some of the most important water issues that currently relate to tensions between governments in Central Asia and create everyday hardships and social discontent are: hydroelectric dam development, irrigation management, infrastructure failures, tensions about water sharing along militarized border zones, and flooding potential. Some of these problems result from the rapid post-Soviet collapse transition to the current state boundaries without parallel transformation of water resource infrastructure and energy production. Additionally, socio-economic pressures in border regions and these countries more broadly – especially in Kyrgyzstan and Tajikistan – create livelihood challenges that are and will worsen with unpredictable weather, precipitation, and temperatures. The collapse of the Aral Sea ecosystem is a significant livelihood threat. The water-energy nexus is an important sphere to consider the potential for deepening disputes between states but also contentious politics within countries.

In the future the biggest physical risk to water supplies and thus contention about transboundary waterways is glacier loss and precipitation changes. Migration pressures from these water regime shifts are a possible link between water and conflict. Migration is already a significant socio-economic pressure; for example, close to 20-25% of Tajikistan and Kyrgyzstan's populations work abroad. Adaptive capacity matters for how governments deal with the everyday issues citizens face in confronting acute water challenges, such as drought, flooding, electricity outages, and extreme weather events. Kyrgyzstan and Tajikistan currently face significant capacity limitations in this regard, and all five countries have vulnerable rural places and populations.

There are ongoing disagreements between Central Asian governments about the key rivers in the Aral Sea watershed - the Syr Dar'ya and Amu Dar'ya (see map 2), and annually recurring seasonal tension between communities in securitized border regions such as in the Ferghana Valley (where Kyrgyzstan, Tajikistan, and Uzbekistan meet). But also there exist complex functioning relationships from the local cross-border community to the international level. Cooperation is more regular than disagreements and tense events, although at times that cooperation seems fragile. Shared 'upstream-downstream' waterways, such as these Aral basin rivers, are more difficult to coordinate and create opportunities for heightened rhetoric, also depending on how military and economic power is distributed. (In contrast, cooperation is achieved more easily among states whose boundaries were created along a shared river, such as the Central Asian Chui-Talas, Panj, and Khorogos rivers). Whether or not this relationship is more cooperative or more conflictual depends in part on how – at any given time – a country's government identifies water as integral to the nation-state's survival, as part of the nation-building project, or as leverage in other relationships with neighboring countries.

In sum, the biggest conflict risks in Central Asia are political and economic: government willingness to tackle people's everyday struggles with water and power, subsequent contention between people and governments about non-responsive policies, and regional leaders' use of nationalistic rhetoric to lay claim to waterways and rationalize particular waterway uses.

Central Asia – the Hydrosphere

The Syr Dar'ya and Amu Dar'ya terminate in the former Aral Sea, in a depression that straddles Kazakhstan at its north and northwest end, Uzbekistan on its east and south, and Turkmenistan to the south and west. These rivers are primarily glacier-fed, from glacier fields in the Tien Shan mountains which the Kyrgyzstan-Kazakhstan-China borders transect and the Pamir-Alai mountains where the Tajikistan-Kyrgyzstan border meets and mountain ranges join in Tajikistan and Afghanistan (see map 1). (According to NASA, the average runoff from glaciers in Kyrgyzstan is 15% of the total, higher during warmer seasons.) The glaciers in both the Pamirs and the Tien Shan have recorded on average steady declines in the 20th and early 21st centuries. The smaller glaciers at lower elevation – those closest to and most important freshwater source for populated areas - are primarily at risk and receding fastest.

The Aral Sea – a terminal lake in the Kyzyl Kum desert, once the fourth largest freshwater lake in the world– is infamous as what the United Nations called the worst environmental disaster of the 20th century. Unsustainable irrigation development upstream mostly for cotton production – canals and irrigation network diversions – siphoned off enough water to desiccate more than 90% of the lake area from 1960-2014 (see images 1 and 2). The lake split in two in 1987 and subsequently into four parts (see image 3). As of this summer 2014, NASA satellite images now show separate seas, and that the largest portion of the lake no longer exists as such (see image 4). The average annual inflows have changed from 55 km^3 in 1910-1960 to 4.1 km^3 in 2006-2010. Water salinity levels went from 10 percent in 1960 to more than 100 percent salinity in the West and East Basins of the Large Aral. In August 2005, with World Bank funding, the government of Kazakhstan constructed the 8-mile Kok-Aral dam, further reducing flows to the Large (southern) Aral and holding the Syr Dar'ya flow in the Small (northern) Aral, which increased to 22.0 km^3 in 2011. This dam construction has been identified by the World Bank as a success story, but it contributed to a faster decline of the majority of this lake ecosystem, once one of the largest in the world, which is now gone and irretrievable. Thus the Kok-Aral dam is an example of an international intervention that arguably provided localized development benefits but negative ecosystem impacts, thus serves as a warning of caution for the ways in which the international community engages in regional water issues.

There are numerous environmental, health, and economic ramifications of the loss of the Aral Sea ecosystem, such as a shortened growing season, water salinity increase, desertification, dust storms (see images 5 and 6), soil salinization and water logging, species loss, high rates of skin disorders, esophageal and other cancers, and a myriad of other health issues among residents from poor air and water quality interacting with nutritional limitations. Restoration efforts now focus on the Small Aral (in Kazakhstan) and the former deltas of the Amu Dar'ya and Syr Dar'ya. The Large Aral is mostly abandoned – the Eastern Basin seems unrecoverable and the deeper Western basin is in decline. This desiccation has led to a significant warming of 6 degrees C (2m) air temperatures over the lake bed and 1 degree C in 200 km area around the former lake boundary.

The Aral Sea matters for many ecological, socio-economic, cultural, and political reasons. In terms of conflict, it is an indirect factor, as further declines in quality of life among residents – mostly in Uzbekistan where communities are closest to the now disappeared Large Aral – could contribute to animosity toward the government and dissent. Most important are continuing problems of access to freshwater, soil quality loss, and reductions in the length of the growing season. A long growing season is crucial for community-scale food production, as well as for the cotton sector which the state tightly controls and from which it draws revenue. Water diversions – such as in the Qaraqum Canal (see image 7, Hanhowuz Reservoir, Turkmenistan) – primarily for cotton production were a main cause of the Aral Sea's collapse and ironically cotton production is impacted by this ecosystem loss.

The Aral Sea crisis has also figured into regional narratives about the water-energy nexus, and criticisms of Kyrgyzstan's and Tajikistan's upstream hydroelectricity development plans by Uzbekistan and Kazakhstan. In 2007, Uzbekistan's President Karimov stated that the construction of large dams upstream would "speed up the ecological catastrophe of the desiccation of the Aral Sea and make it practically impossible to live for tens of millions of residents of Kazakhstan, Turkmenistan, and Uzbekistan."

This rhetoric rings false as the government of Uzbekistan has done very little to address the main causes of the catastrophe, but rather has impeded civil society attempts to draw attention to the Aral Sea disaster and the problems with cotton monoculture – at times imprisoning critics of its environmental policies – consistently blocks regional action to tackle the Aral Sea issue, does not allow the government of the Autonomous Republic of Karakalpakstan (the region most impacted by the Aral Sea's collapse) to practice its political autonomy, and has resisted change in its agricultural sector. However, claims of threat to the Aral Sea from upstream hydroelectricity development cannot be dismissed outright because dams do create river regime changes and there is inadequate public (national) and international level discussion of the ecological impacts of these dams' construction.

Hydroelectric Dam Development

One point of contention at the inter-governmental level surrounds the issue of hydroenergy development in the upstream countries, Kyrgyzstan and Tajikistan. Downstream countries are concerned that these developments will alter the waterways, impacting availability and timing of flow for communities that rely on this water for irrigation and flood control. The key issue here is how the timing of releases is managed. Everyday experiences in near border regions are important locations of ongoing tension as well as required cooperation. Water resource concerns are widely shared across the region. In Kyrgyzstan, for example, I conducted a nation-wide public opinion poll in 2009 in which water ranked highest among environmental concerns in every single province; 48.41 percent of Kyrgyzstanis were concerned most about water supply or pollution. Water supply and pollution also figured first for the majority of 140 key informants I interviewed from 2009-2013. Water infrastructure maintenance is a major problem throughout the region and a consideration for international assistance.

Kyrgyzstan and Tajikistan rely on hydropower for the vast majority of their electricity production (~90% in both countries). Both countries are dependent on Soviet-era dam infrastructure for this energy supply, since the trading relationship of importing coal, oil, and gas from Kazakhstan, Turkmenistan and Uzbekistan was monetized at the Soviet break-up and the upstream countries do not have adequate fossil fuels for their energy demands, especially in the cold winters which these mountainous countries experience. Given this predicament, both Kyrgyzstan and Tajikistan are reviving Soviet-era plans to upgrade and expand the hydroenergy sector, primarily by constructing a series of dams. Kyrgyzstan is constructing the large Kambar-Ata 1 and 2 dams and four smaller dams, all on the Naryn River, a major tributary of the Syr-Dar'ya, which runs through the country and into Kazakhstan and Uzbekistan. The Kambar-Ata 2 dam is producing electricity already, and Russian state investment in the upper dam construction was announced in 2013. These four small dams are targeted for completion in 2019. Tajikistan is investing heavily in the Rogun dam which has generated the most ire from Uzbekistan's government. The government of Uzbekistan has seemingly engaged in economic retribution through border securitization and cutting off gas supplies to both countries at various times.

In September 2012, during an official visit to Astana, Uzbekistan's President Islam Karimov noted that leaders of upstream Kyrgyzstan and Tajikistan forget the transboundary nature of the region's rivers. "Because today many experts declare that water resources could tomorrow become a problem around which relations deteriorate, and not only in our region. Everything can be so aggravated that this can spark not simply serious confrontation but even wars." Kazakhstan's President Nursultan Nazarbayev then stated, "To our neighbors and brothers who are 'sitting' on the upper reaches of these rivers, we send another 'fraternal signal' that we -- Kazakhstan and Uzbekistan on the Amu-Darya and Turkmenistan, located downstream -- most of all perceive the shortage of water; each person feels it, because this is their life; this is the life of millions of people." One year later when these two leaders met in Tashkent (June 2013), they called for UN review of Kyrgyzstan's and Tajikistan's dam construction plans and stated their common position on resolving regional water and energy problems through international law. President Nazarbayev noted, "We want to send a friendly message to our neighbors that we ourselves need to address these issues together. There are no unsolvable problems and issues. ... We are ready to meet you halfway. Kyrgyzstan and Tajikistan have transport and energy issues. We are also prepared to deal with these issues." So downstream states continue to pressure upstream states to halt large dam construction or open the process to neutral evaluation and regional discussion.

However, both Kyrgyzstan and Tajikistan also face the reality of electricity shortages and concerns about the impact of power outages on their populations' well-being and the possibilities for popular discontent. For example, in some parts of Tajikistan, electricity is available only three hours a day. In 2010, water sector partial privatization and doubling of electricity, water, and heating tariffs helped trigger protests against the Kyrgyz government. This rate increase was widely unpopular in light of the daily electricity shortage most Kyrgyzstanis experienced in 2008-09 (71.4% in my survey at the time indicated experiencing significant blackout impacts). When government forces cracked down on protesters on April 7th, President Bakiev was run out of power. Since this moment, the interim government of Roza Otunbaeva and current government of Almazbek Atambayev have cautiously approached the electricity tariff issue, invested heavily in the energy sector, and moved forward on Bakiev-era plans to develop hydroelectricity dams and invest in grid improvements. However, tariff policy was again broached this year, as water levels are again low in Toktogul dam and blackouts this winter are imminent. In this context, it is hard to imagine Kyrgyzstan halting dam construction plans.

This discussion demonstrates the many ways water is seen and valued by people living in the region, beyond water as 'an emerging threat'. Threat rhetoric has political meaning and is used by governments to mobilize citizens. Therefore, the U.S. Government should be cautious when using threat and risk language in our search to understand potential water problems and in our approach to assisting with these issues, as this could contribute to an at times difficult dialogue between Central Asia states.

Despite what seems like widespread support for hydroelectricity development to address these severe energy needs, dams – especially large ones with high numbers of people displaced – are often the sites of social contestation and major social disruption. The existing major dams in Kyrgyzstan and Tajikistan are respectively Toktogul (the largest reservoir in Central Asia) and Nurek (the highest dam in the world, see image 8). Historically there was contention surrounding the construction of these two large dams, mostly regarding displacement, and plans to build the Rogun may have also been a factor in the Tajik civil war. In Kyrgyzstan, most of the dams currently under construction or planned will not displace large numbers of communities and residents (although there still may be contentious issues in those communities affected or due to loss of communal pasture lands), but the Rogun dam is estimated to displace more than 40,000 people and is more unpopular than Kyrgyzstan's dams where there is little criticism.

Both countries have embarked on public relations campaigns to paint these hydroelectric development projects as fundamentally crucial nation-building endeavors, identifying environmental and energy security and regional independence goals. For Tajikistan it seems that the current campaign to show Rogun in this light is important for garnering adequate popular support for a project that will potentially negatively impact so many. So nationalistic rhetoric by the leaders of both countries – in Kyrgyzstan this rhetoric about the Naryn river dams spans several administrations – is used to mobilize internal support but contributes to regional tensions by signaling the domestic political importance of these projects regardless of other countries' concerns.

In response to some of these pressures and to attract international financing, the government of Tajikistan asked the World Bank to conduct an impact assessment of the Rogun dam, which was completed in September 2014. The report concluded that this is a technically feasible project, and identified the need for better population relocation plans, limiting downstream impacts during the reservoir filling period, and managing seasonal releases of water to minimize disruption to available flows and to ensure flooding is managed (this is a concern for Kazakhstan along the Syr-Dar'ya – see image 9 of flooding along that river basin in 2004). There is ambiguity in the regional water sharing system and treaties, with no clear language about seasonal transfers and annual flow allocations determined in semi-annual meetings, so clarity is needed about how Rogun will be operated, or tensions will remain and perhaps increase during dam construction until the filling period begins and flow changes become evident. Another point of contention is about the CASA-1000 project to export electricity from Kyrgyzstan and Tajikistan to Afghanistan and Pakistan. Uzbekistan opposes this project as it arguably depends on construction of the Naryn dams and Rogun for adequate electricity production to export.

Limitations of Regional Water Institutions

Regional water institutions in Central Asia are elaborate, but they are widely regarded as weak and based on vague treaties determined by Soviet era considerations. These institutions work at the Aral basin regional level, rather than one for each major river basin. The International Crisis Group (ICG) recently suggested improving the institutional framework by creating multiple bilateral agreements instead of the current dysfunctional set of multilateral agreements. Kazakhstan and China recently created regional waterway institutions to help resolve water sharing and pollution concerns, yet the substantive outcomes are still to be seen. One somewhat successful example of a cooperative water sharing body was the Chui-Talas Rivers Commission between Kazakhstan and Kyrgyzstan. This commission facilitated resolution of a number of issues, including payments for upkeep of water supply and storage infrastructure on the Kyrgyzstani side and improved monitoring for water forecasts. However, this relationship was tested in 2010 when Kyrgyzstan closed off canals into Kazakhstan during the growing season in the summer. Kazakhstan had closed the borders following violence in Osh, Kyrgyzstan in June 2010, and it seems the stoppage of water flow was a response; Kazakhstan immediately re-opened the borders. This is a dangerous example of water used as a weapon. It does not mean conflict potential has gone up markedly, but it does reveal the limitations of regional institutions.

Water Pollution Concerns

Often forgotten in discussion of scarcity or distribution are issues of water pollution, which create immediate impacts on residents relying on these sources but also interact with reduced supply to effectively create more scarcity, where water becomes unusable. Non-point water pollution problems include inadequate sanitation and agricultural chemical and sediment runoff, including high concentrations of DDT. Industrial, point-source pollutants include aging Soviet-era industrial facilities like the Khaidarkan mercury plant (the last exporting mercury mine in the world) in Kadamjay, southern Kyrgyzstan and the TALCO (Tajik Aluminum Company) aluminum factory in Tursunzoda, Tajikistan. There are numerous radioactive uranium tailings and waste sites throughout the region – such as Mailuu-Suu and Min-Kush, Kyrgyzstan in the Naryn River basin – and oil pollution around the Caspian Sea, such as in the Karachaganak oil field, Kazakhstan and Turkmenbashi Bay, Turkmenistan. Pollution concerns - combined with increased extraction for a growing population and expanded cotton production - are central to the Kazakhstan-China relationship in the Ili and Irtysh rivers (the largest and most contentious among the 20 transboundary rivers between them, see map 2), as industrial development rapidly grows in China's Xinjiang Province (see image 10).

Emerging Concerns – Glacier Decline & Permafrost Changes

Much like in California, water planners in Central Asia have long been concerned with the implications of climate change for this glacier-dependent (glacio-nival) regional hydrological regime. Projections of decreases in Tien Shan Mountain glacier surface areas and volume over the next few decades means the region would shift from this hydrological regime to a pluvio-nival one, which is precipitation dependent. This means greater variation in water levels between seasons and between years. The Intergovernmental Panel on Climate Change (IPCC) prediction for Central Asia by 2050 is a 4-8% increase in winter precipitation and decrease in summer precipitation by 4-7%. There are ways of adapting to this slow-moving process, but this requires significant international support. The latest IPCC report identified a more immediate hazard of breach potential for moraine-dammed glacial lakes – such as Lake Petrov in Kyrgyzstan and a number of naturally dammed lakes in the Pamir mountains, Tajikistan.

Another consideration for the Tien Shan mountains that comes out of the most recent IPCC report is impact of climate change on melting permafrost, which is now expected sooner than previously predicted. The location of Kyrgyzstan's Kumtor gold mine in a glacial zone in the Ak Shirak range of the Tien Shan mountains – it is the only mine in the world operating on active glaciers, including mining the glaciers – connects with both glacier decline but also waste management with risk for water quality. The Kumtor tailing pond is situated between Lake Petrov and the Kumtor River, a tributary of the Naryn River. The unlined tailing pit holds more than 34 million m^3 of wastewater and tailings from the cyanide leachate and other chemicals used to process gold at Kumtor and relies on underlying permafrost for continuous, permanent, impermeable containment. Therefore, changes in the permafrost underneath this extensive tailing pit at the headwaters to the Naryn river and breach threats to Lake Petrov above the tailing pond are concerns that should be monitored.

Policy Implications: What should the U.S. Government be doing?

The policy implications of the water security situation in Central Asia are five-fold, ranging from broad considerations to specific intervention opportunities. First, given the tight relationship between water and energy, regional disputes related to the distribution of both, and the problem of energy insecurity at the household level, it is most important to provide strong support for investment in and expansion of renewable energy sources and off-grid household energy systems. Examples include micro-hydro, solar, wind power, and geothermal energy source developments. It is also possible to invest in the technological development of these sectors in the region, which would provide longer-term benefits than short-term pilot or small-scale projects, and in national grid expansion and improvements. In light of the political, social, and environmental problems of both fossil fuel and hydroelectric development, renewables is the energy sector which the international community and the U.S. government should seek to expand.

Second, one of the key points of tension for countries in the region is predicting water flows, glacial decline, and permafrost changes. Supporting scientific glaciology and hydrology research and monitoring in Central Asia would be a valuable, lasting contribution to a once vibrant sector that is now drastically underfunded. In Kyrgyzstan and Tajikistan, the need is particularly acute for glacier monitoring instrumentation, which would be useful for all Aral Sea basin states. There are currently some efforts to assist and support in improving this instrumentation and regular monitoring. One example is the German funded Regional Research Network "Central Asian Water" CAWa, which works in partnership with the University of Idaho. Another is the World Bank Central Asia Hydrometeorology Modernization Program (CAHMP). However, scientific research needs to be funded to a greater extent and continuously. This is a funding area that would provide regional benefit in predictability and information transparency.

Third, water pollution is often missed in discussions of conflict, but it is fundamental to tackle everyday problems residents face and consequentially is valuable for maintaining political stability. The U.S. Government could increase significantly its contributions to international efforts to aid in sanitation improvements and tackle other pernicious pollution problems, providing greater support to local and international organizations involved in this effort, such as the Central Asian Alliance for Water. USG funding to help address the uranium tailings legacy would be positively received by multiple countries, upstream and downstream, and if done carefully and in the long-term, would be welcomed by affected communities. Support for tackling these pollution problems would contribute to water supply improvements and help address serious health impacts.

Fourth, one of the most important roles that the U.S. Government has played over the last few decades has been funding for US scholars studying the region. The continuation of support is necessary for understanding emerging issues in Central Asia such as water security concerns we are discussing today. I sit before you as a product of multiple research grants through Title VIII funding organizations such as ACCELS/American Councils, IREX, and NCEEER. Without continuous support for research, the ability of this committee to understand the region will be affected and limited. Removing this link between our community of scholars and the community of scholars and residents in the countries we study would be shortsighted. I encourage continued support for these research funds as a tangible U.S. government involvement in Central Asia that produces observable results and positive transnational relationships. Similarly, support for educational institutions and programs in the region, such as the American University of Central Asia, where I taught in 2001, is a similarly vital contribution.

Fifth, some of the main fears among people and planners in the region about climate change relate to glacial decline. Thus, one of the most important ways the U.S. Government could help mitigate future water supply problems in Central Asia is to begin tackling this problem. The U.S. Congress should recognize the important role of climate change in creating ecological problems for human communities around the world, particularly in vulnerable places currently less economically capable of adapting to these shifts, like many rural areas in Central Asia. Some of those changes are already discernible, and the patterns identified as probably or likely – such as glacier loss – will have clear impacts on water distribution in the region. Recognizing this important role means seeking to tackle the overall issue of climate change by reducing the U.S. contribution to greenhouse gases over the next dozen years and decades, as well as targeting aid in Central Asia to mitigation and adaptation efforts.

Thank you, I look forward to your questions.

Mr. ROHRABACHER. You are next, Doctor.

STATEMENT OF KATHLEEN KUEHNAST, PH.D., DIRECTOR, CENTER FOR GENDER & PEACEBUILDING, UNITED STATES INSTITUTE OF PEACE

Ms. KUEHNAST. Thank you so much, Chairman Rohrabacher, Ranking Member Keating, Mr. Meeks and other members of the House Foreign Affairs Subcommittee on Europe, Eurasia and Emerging Threats.

Please note that the views I express today are solely my own and not those of the U.S. Institute of Peace, which does not take policy positions.

I want to emphasize three points related to Central Asia and water issues. Water is a mismanaged resource, not necessarily a scarce one. Central Asia is not a friendly neighborhood for water management.

Given these predicaments, we can ask why there hasn't been more conflict over water in the past and how this might change in the future.

According to World Bank estimates, only 21 percent of water in Central Asia is used effectively. A recent report in the journal Nature found that on average a person in Turkmenistan consumes four times more water than an average American and 13 times more than a Chinese citizen.

Certainly, the Soviet centralized state enforced the rules for water allocation among the republics, regulating and maintaining canals, pumping stations, irrigation facilities, dams and reservoirs.

The post-Soviet era, however, brought the creation of five Central Asian states, resulting in a predicament where 98 percent of Turkmenistan's water supply and 91 percent of Uzbekistan's originates outside their borders.

While poor water management and wasteful practices are core issues in Central Asia, the factors that have kept the regional tensions over water and energy resources from spilling over may no longer hold.

Consider the demographic challenge of Central Asia. Roughly half the population is under 30 years of age and most of these young people are worse off than their parents' generation, with higher rates of illiteracy, unemployment and poor health.

Inattention to the needs of this disenfranchised age group increases the risk of local level conflicts in the individual countries and throughout the region, especially with known extremist groups infiltrating these countries.

But also add a widening gap and tension between elites and the poor, weak governance along with the prevalence of patron-client relationships, loyalty, manipulation of formal rules.

Add an increasing fear among the local populations that water and energy problems will be resolved at the end of a gun, especially as the number of small arms surge, coming up from Afghanistan.

Add the fact of the large number of labor migrants from Uzbekistan, Kyrgyzstan and Tajikistan leaving empty villages except for their labor widows, as they are often called. We have a security problem.

The recent prediction that Central Asia will see slower growth this next year due to the economic stagnation of countries outside the region, and finally, President Karimov's exit, whenever it happens, there will be a change in the dynamics of the region.

His autocratic and oppressive governance has held the border tensions in check. However, new leadership may bring uncertainties with regards to containing hostilities.

In summary, I would like to highlight several points for the U.S. Government including Congress to consider. Support good water data collection and share information with consumers, farmers, businesses and policy makers alike.

Engage young people through small grants and encourage entrepreneurial social marketing and research studies to address overuse of water by fellow citizens. Teach critical problem solving skills in U.S. training and educational exchange programs with Central Asian youth.

Engage women and civil society as women are often at the nexus of daily water management. And finally, ensure effective conflict management skills at the local and national levels. This is where the U.S. can most effectively contribute by education and training that offers skill-based approaches to negotiation and conflict management.

The U.S. Institute of Peace considers conflict a normal condition of human societies. However, much can be done to prevent violent conflict from being the default mechanism for solving the problem at hand.

Central Asia deserves American support for conflict prevention. Thank you. I am happy to answer your questions.

[The prepared statement of Ms. Kuehnast follows:]

United States Institute of Peace

. . .

An independent institution established by Congress to strengthen the nation's capacity
to promote peaceful resolution to international conflicts

. . .

"Water Sharing Conflicts and the Threat to International Peace"

Testimony before the House Foreign Affairs

Subcommittee on Europe, Eurasia and Emerging Threats

U.S. House of Representatives

Kathleen Kuehnast

United States Institute of Peace

November 18, 2014

Good afternoon and thank you to the House Foreign Affairs Subcommittee on Europe, Eurasia and Emerging Threats Chairman Rohrabacher, Ranking Member Keating, and other members of the Subcommittee for this opportunity to testify before you today.

My name is Dr. Kathleen Kuehnast. I am a socio-cultural anthropologist and an expert on the Central Asia country of Kyrgyzstan, where for over a decade I worked as a social scientist for the World Bank, focusing regionally on Central Asia and thematically on conflict prevention. My comments here today derive partially from various studies I did in that capacity on the topic of water and border management, and more recently as director for gender and peacebuilding at the United States Institute of Peace (USIP). I have recently published a USIP report on the potential for violent extremism in Kyrgyzstan. The U.S. Congress created the Institute 30 years ago with a mandate to prevent, mitigate and resolve violent conflicts around the world. The Institute does so by engaging directly in conflict zones and by providing analysis, education and resources to those working for peace. USIP experts work on the ground in some of the world's most volatile regions, collaborating with U.S. government agencies, non-governmental organizations, and local communities to foster peace and stability.

The views I express today are solely my own and do not represent those of the United States Institute of Peace, which does not take policy positions.

This is an important hearing on how disputes over water sharing or water management contribute to regional hostilities. My comments will focus on the everyday use and contestation of water in Central Asia between the upstream countries -- Tajikistan and Kyrgyzstan and the downstream countries—Uzbekistan and Turkmenistan. Since the early 1990's, the Soviet system of management has fallen apart and now water is a major source of tension and instability. I also offer details of these situations and present some policy considerations for how the U.S. government, including Congress, can help diffuse and resolve these impasses.

I will emphasize three points related to Central Asia and water issues:

1. Water is mismanaged resource, not necessarily a scarce one.
2. Central Asia is not a friendly neighborhood for water management.
3. Given these predicaments, we can ask why hasn't there been more conflict over water in the past, and how might this change in the future.

Water is wasted.

Water is wasted daily as the result of its Soviet-era unsustainable water practices, deteriorating infrastructure and two decades of state-centric power struggles, mismanagement and corruption.

According to World Bank estimates from 2005, water use in Central Asia is as high as 12,900 cubic meters per hectare, but only 21 percent of this is used effectively. The remaining 79 percent is lost to unlined canals and irrigation inefficiencies [UNEP, Environment and Security: Transforming Risks into Cooperation. 2005]. This compares with roughly 60 percent loss in developing countries. A recent report in the science journal *Nature* found that on average, a person in Turkmenistan consumes four times more water than an average American, and 13 times more than a Chinese citizen [NATURE, VOL 514. 2 OCTOBER 2014].

Access to water is one of the most contentious local issues in Central Asia today. The competition for water is fierce, particularly in the densely populated Fergana Valley, split between Kyrgyzstan, Tajikistan, and Uzbekistan. The valley has some of the most extensive irrigation systems in the world, serving approximately 22 million people dependent on irrigated agriculture.

The Soviet regime left an integrated water system of dams, pumping stations and canals for irrigation and energy generation. The centralized state enforced clear rules for water allocation among the Central Asian republics in order to support extensive cotton production for the entire USSR. Top-down Soviet water management may have minimized conflict, however, it intensely ignored the environmental, social and economic aspects of natural resource management [Kuehnast et al. WB 2008].

With aspirations of creating an oasis in the deserts of Central Asia, Soviet practices from the 1950s onward included growing cotton, rice, and potatoes, all of which are water intensive crops and some are particularly damaging to the soil, which I would add cannot be left out of the water management equation. In the post-Soviet era, the individual countries have recalibrated their policies and behaviors following the termination of Soviet-styled subsidies. The United Nations Environment Programme [2005] estimated that the creation of the Central Asian states in 1991 meant that 98% of Turkmenistan's water supply and 91% of Uzbekistan's originate outside their borders.

Without a central coordinator and enforcer, it is difficult, if not impossible, to maintain canals, pumping stations, irrigation facilities, dams and reservoirs in the region in a way that promotes efficient water management. For instance, the water-energy nexus evolved in the neighborhood of Central Asia, where the upstream countries, dependent on insufficient hydro power, especially in winter, were then traded gas-powered electricity generated in Uzbekistan and Turkmenistan at subsidized prices. This arrangement allowed the upstream countries to meet their winter energy needs and therefore conserve their water reserves, which would be released for irrigation in the downstream countries during the summer. The breakdown of the subsidy regime after the collapse of the Soviet Union led upstream countries to release water in winter to generate power, meaning that less of it was available during the spring and summer for irrigation.

This is not a friendly neighborhood.

Many local people feel that the roots of the water crisis rest in the lack of comprehensive, coordinated decision making by Central Asian governments. On the other hand, even when inter-governmental agreements are made, they are often not enforced locally.

At the heart of this competitive, one-upmanship is the problem of the upstream countries of Kyrgyzstan and Tajikistan with a combined population of about 10 million, wanting to build large hydropower plants that need huge reservoirs to meet their energy needs in the absence of an integrated energy and water system. Downstream is Uzbekistan with over 27 million people, wanting the same water for irrigation of crops in the spring and summer. Instead the result has been for many in the Fergana Valley rising ground water and waterlogging, both leaving a massive rodent problem, along with disease.

Why hasn't there been more conflict?

The international community should take into account the historical and socio-cultural practices that have prevented violence and conflict. Two decades after the collapse of the Soviet Union, the factors that have kept the regional tensions from spilling over may no longer hold.

Experts have repeatedly predicted that interethnic tensions could precipitate more major conflicts in Central Asia. In spite of these proclamations about interethnic tensions, World Bank research [Kuehnast et al, WB 2008] suggests that such tensions have emerged more as outcomes rather than drivers of local conflicts. In addition, the widening gap between the elites and vast numbers of the poor, weak governance and conflicting rule systems, in a context of economic uncertainty, high unemployment (especially among youth), and criminalization of the economy are creating the context for local conflicts.

Another critical point to consider is that there has been no political regime change for the last two decades in Uzbekistan. President Karimov's exit—whenever it happens—will change the dynamics in the region. His autocratic and oppressive governance has held the border tensions in check. However, new leadership may bring uncertainties with regards to containing hostilities. Concerns about the number of small and large arms flowing in from Afghanistan have both Uzbeks and Tajiks worried about how issues might get resolved. A recent USIP study [Zenn, Kuehnast. 2014] found people are afraid problems will be resolved at the local level at the end of a gun.

Local conflicts must be also considered in the context of clan and business networks that form the basis of local power on which national leaders depend. The prevalence of patron-client relationships, connections, loyalty, manipulation of formal rules and force are the key parameters for governance in these states. At the very core of water management, conflict management depends on these clan and business networks.

The gap between the poorly enforced formal, top-down regulations and the local, customary laws that rely on social influence, cohesion and respect for elders has created a situation in which discordant "legal pluralism" has emerged. Therefore, individuals must constantly contend with the issue of "whose rules rule" in everyday contestations. The clash between various rules is intensified at neighboring borders, where differing policies collide among these not-so-friendly states. In many instances, even the formal border demarcation is contested, leaving a no-man's land that exacerbates local tensions and conflicts.

Agriculture is the key to the survival for many in the region, and in some areas, competition for land and water is intense. While wasteful practices and poor water management are core issues, one must also consider the demographic challenge of Central Asia: roughly half of the population is under 30 years of age, and most of these young people are worse off than their parents' generation. As a group, this generation of youth has higher rates of illiteracy, unemployment, poor health and drug use than other age groups. Inattention to the needs of this disenfranchised group significantly increases the risk of local level conflicts in individual countries and throughout the region. In addition, water and border conflicts are intertwined with other issues, and any crisis in the region will not be a stand-alone water conflict, but incorporate elements of violent extremism (including the influence of ISIL, to which the Central Asian jihadist group IMU has pledged its support), drug networks, and the yet unknown impacts of the Afghan transition, political nationalism and Russian trade union pressures [Zenn, Kuehnast, 2014].

Recommendations

What policies and practices should the USG and international community emphasize in order to change the trajectory of water conflicts? Some of the key points I'd like to highlight for the U.S. government, including Congress, to consider:

- The U.S. needs to be support good data collection and accurate information communication to be shared across the five countries. Transparency is critical for the consumers, the farmers, and the policy makers alike. Community engagement as well as linking regional and academic leaders in dialogues can establish forums for long-term effective policy framework and initiatives.

- U.S. government and international policies should engage young people through small grants that encourage entrepreneurial marketing and research studies to address the issue of the overuse of water by fellow citizens in both urban and rural areas.

- Critical problem solving skills are essential and need to be a part of US training programs and educational exchange programs. Technical education and business acumen are "musts" for the future.

- Finally, more emphasis needs to be placed on learning effective conflict management skills at the local and national levels. This is where the United States can most effectively contribute by education and training that offers skill-based approaches to negotiation and conflict management.

The U.S. Institute of Peace considers conflict a normal condition of human societies, however much can be done to prevent violent conflict from being the default mechanism for solving the problem at hand. Central Asia deserves American support for conflict prevention. Thank you. I am happy to answer your questions.

The views expressed in this testimony are those of the author and not the U.S. Institute of Peace, which does not take policy positions.

Mr. ROHRABACHER. Well, thank you very much to all three of you for your testimony. Some of the statistics that you were quoting both about Egypt's use of water and both Tajikistan and Uzbekistan's use of water that is coming from the outside of their country is just a bit overwhelming from countries like the United States that have so many of our own resources.

Let me note, in California I am not sure exactly what percentage of California water comes from the outside, which will lead me to a question about, I guess, is desal being used at all, Dr. Sullivan, in Egypt?

Mr. SULLIVAN. It is being used in the Sinai for some of the hotels and for some of the military bases and also on the north coast.

But part of the problem is that about 98 percent of the population of Egypt lives on the Nile and the Nile stretches all the way to the Sudan from the Mediterranean and you can't desalinate the Nile.

What you would have to do is have huge pipeline systems set up to bring the desalinated water from either the Mediterranean or from the Red Sea or from other parts of the——

Mr. ROHRABACHER. Well, the—isn't Cairo near the ocean?

Mr. SULLIVAN. Well, it is actually many miles from the ocean.

Mr. ROHRABACHER. That is interesting. Yes.

Mr. SULLIVAN. It is a good two and a half hour drive to get to Alexandria from Cairo, and for that water to move from the Mediterranean Sea to Cairo would be extremely expensive infrastructurally and also it takes a lot of energy.

If they set up solar desalinization plants that may be a better way of doing things. They also—they already have a shortage of gas. They are a net importer of oil.

A country that, when I was living there in the 1990s, was a country that was hoping for great exports of oil and natural gas.

Mr. ROHRABACHER. So the—what we are talking about is projecting what could be but what isn't and I would say my own observation of how close things are or how far things are away it all depends on your perspective.

Being 2 hours from the ocean may or may not be a long way for certain societies.

Mr. SULLIVAN. Well, in L.A. that is just across the corner.

Mr. ROHRABACHER. That is right.

Mr. SULLIVAN. But in Cairo the traffic is difficult but I am talking about on a Friday morning during an Eid. It is a long distance.

Mr. ROHRABACHER. Mm-hmm. We actually get our water from— a lot of water from the Colorado River but we also now are—have a—put a—made a lot of large investment in developing new desal technology, which may be—have an impact on the Egyptian problem because at least it could provide some areas that have some access to the—to the ocean or the Mediterranean Sea or whatever, the Red Sea as well, a means of having fresh water because the price of desal is going down.

I am on the Science Committee and I can tell you that the new invention by Lockheed of use of graphene in your system will bring down the cost—of energy cost of desal and thus will bring down the cost dramatically.

Whether or not how long that takes to put in place, that type of desal project and with the energy level that is still required by that may or may not come in time.

Let me—is there anyone right now involved with trying to mediate or arbitrate a difference between Ethiopia and Egypt?

Mr. SULLIVAN. Well, the President al-Sisi has had many discussions with the Ethiopians. There are conciliatory moves including some trade deals.

The rhetoric has been calmed down, certainly since the regime of President Morsi and the Muslim Brotherhood which, by the way, if water stress comes in an extreme manner to Egypt there is a very good chance the Brotherhood will come back because it will be a failure for General al-Sisi—President al-Sisi. There are many people trying to find some middle ground here.

Mr. ROHRABACHER. Who—what—isn't there—is there any international organization we have now that we can call upon to do that?

Mr. SULLIVAN. Well, there is the Nile Basin Initiative and that——

Mr. ROHRABACHER. Right. Seeking a—to mediate or to arbitrate this dispute or is it just there to try to facilitate communication?

Mr. SULLIVAN. Facilitate communication. It is also rather toothless on this one. When you have two countries that are at odds it is very difficult to find a middle ground.

The United States may act as a convener in some place distant from their two countries.

Mr. ROHRABACHER. So you think the United States—that we could step forward and offer both of these countries a—to play a mediation role or an arbitration role?

Mr. SULLIVAN. We could. We could, and it also may be a way to repair some of the damage to U.S.-Egyptian relations that occurred with the cutting back of military aid in a time when they needed it the most and then they turned to Russia, which right now seems to be a problematic country for the United States.

Mr. ROHRABACHER. That happened with a dam in the past, didn't it?

Mr. SULLIVAN. Yes, it did. Yes, it did. That would be the Aswan High Dam. But desalinization, getting back here, they are not that—that doesn't seem to me to be sufficient and would take time to ramp this thing up.

Mr. ROHRABACHER. Well, it certainly wouldn't be sufficient in the Central Asian republics because they are not near the ocean at all.

Let us just note that we did convince the Ethiopians at one point to agree to arbitration of a major dispute that they were in with Eritrea, and I—this has happened during the last administration so you know this is not a partisan remark.

But I thought the behavior of our Government in that whole episode was disgraceful and has undermined our ability to be—to arbitrate other disputes in the sense that Ethiopia—the decision of the arbitration went against Ethiopia in their border dispute with Eritrea and we extracted some kind of other deal with them to help us with some sort of defense-related deal and let them off the hook, basically said they didn't have to follow the arbitration which meant that the message to all of Africa was you don't—you better skip out the arbitration because that just doesn't work.

Even the Americans are going to discard it—what the result is. That was very sad. I would hope that we could come up with someone who could help arbitrate between Ethiopia and Egypt on this.

Your testimony is, from what I take from your testimony, is that if the Ethiopians do take a longer period of filling this dam up with water, a long period of time, it may not be harmful.

They could—it is possible that both sides could actually come out of this okay as long as the Ethiopians were filling their dam in a responsible way. Is that correct?

Mr. SULLIVAN. When you are in a country such as Egypt which is using 98 percent of its water, there is only a 2 percent leeway there, and to see only 2 percent reduction is probably a low probability event.

When they are filling this thing up, they want to fill it up as quickly as possible. Yes, there is a possibility for negotiating a slow fill. But in order to not damage Egypt it would have to be over many, many years.

Mr. ROHRABACHER. That should be something that maybe they could be negotiating, perhaps.

Mr. SULLIVAN. Perhaps, but it also leaves Ethiopia in a position of controlling the water tap if at any time it needs to control that water for the next famine, for the next drought, and then Egypt is shut off.

During the 1980s, 1984 in particular, Egypt was damaged from this and there wasn't even a dam there. Egypt is possibly in a much riskier situation now that this spigot is there, which a hydro dam could be. And you add in irrigation it gets even worse.

Mr. ROHRABACHER. I think that we have got to be very sensitive to that—to make sure that that outcome does not happen.

The Egyptian Government right now is pivotal to peace in that region and for us to—the people have to know that the United States—the people of the United States are on their side in making sure that they do not wake up one morning and find the actions of another government, whether it is Ethiopia or whoever, has dramatically negatively impacted on their economic well being.

So today I would just call on our Government—the United States Government—to do what it can to make sure that the Egyptian people are never put in that spot where a decision made in Ethiopia or some other government will economic—bring their economic well being down—the standard of living down that cause suffering among their people. That is unacceptable as an alternative.

Hopefully, the United States—our Government—will take that as a priority and try to get and solve this problem in some sort of role that we can play, which I will leave to you to answer, again, on the second round.

I now—Mr. Keating.

Mr. KEATING. Thank you, Mr. Chairman. I want to compliment the witnesses for the breadth of what they have covered in that short period that they addressed this.

One thing that struck me, let us talk on the demand side for a second because I think it was Dr. Kuehnast who was mentioning the comparison between some Central Asian countries and China, for instance, where the amount consumed per person is 14 times what it is there and it is four times more than in the U.S.

What areas of demand can be addressed to be helpful, if that is the case? I understand some of that is just water mismanagement itself but also what can be done to make those ratios more in line?

Certainly, China has enormous challenges in this area too and they seem to be doing better.

Ms. KUEHNAST. I would clarify that each of these states are so different and difficult to characterize them all as Central Asia. In this case, it was Turkmenistan.

I think you know one of the problems that we are all talking about is agricultural use and there is great demand. You also have a legacy of the Soviet period that wanted to grow rice in this part of the world, that does grow cotton—as you know, takes a lot of water—grows potatoes and wheat. But all of these are a great drain, and what you have simultaneously is countries coming into their own economic processes and they are needing to pivot an entire generation from agricultural practices to service, to business, and that is taking a generation at least because suddenly not having the Soviet system in place for education it is like retooling everybody.

And so some of it is systemic from the Soviet legacy and some of it is how do you help turn a country toward a forward-looking process of business and other kinds of forms of economic development.

Mr. KEATING. You have also touched on emergencies and disasters. Certainly, disasters—natural disasters like earthquakes can cause problems, or terrorist acts can cause problems as well.

What is being done on the international community to reduce the destabilizing effects of this kind of water-related event? Any ideas? Is there any planning on that or are we just sitting there waiting for one of these things to happen? Dr. Sullivan?

Mr. SULLIVAN. There is a lot of talk—talk, talk, talk. Not much action and no enforcement. That is the answer.

Mr. KEATING. So the answer is we are at great risk to those type of disasters and we are not prepared for that. Okay.

Ms. KUEHNAST. I could add to that, if I could.

Mr. KEATING. Yes.

Ms. KUEHNAST. There are a number of programs for dealing with emergency energy needs in Kyrgyzstan and Tajikistan in particular. So the United Nations and World Bank has a number of programs and USAID has been connected too in supporting some of that.

In terms of natural disaster institutional strengthening, in Kyrgyzstan it is actually a pretty interesting case of improvement where the Ministry for Emergency Services actually strengthened and visibly improved its ability to monitor, map and prevent some of those impacts from natural disasters, mostly from landslides and from earthquakes.

Also, there is a good support and implementation from that ministry to build the capacity in local communities and vulnerable locations.

So I would say that that is a really—actually an interesting impact, one of the most valuable impacts of aid in the region and that ministry has worked to coordinate with ministries across the wider region in mapping and monitoring the potential impacts.

Mr. KEATING. I can't let my time go without mentioning global warming, and——

Mr. ROHRABACHER. I knew it.

Mr. KEATING [continuing]. And Dr. Wooden touched on this with glacial decline. But what extent is global warming and climate change?

What extent will they contribute to the potential for conflict over water resources and how can countries—countries—come together to address the larger impact of this phenomenon? Do you have any examples?

Ms. WOODEN. Well, sorry. Go ahead.

Mr. KEATING. No, you go ahead because you mentioned glacial.

Ms. WOODEN. Yes. So I mean, it is interesting. The research is pretty clear that there will be—there has been a decline in the 20th and 21st—early 21st century in glacial coverage.

This will impact water supply. So the regional leaders know that. Everyday people talk about it. This is kind of a known thing there and thinking about adapting is part of the conversation. I think it is an interesting moment for potential cooperation over this.

Right now, it is a source of tension and concern because it is an uncertainty. Exactly what will those precipitation patterns look like, it is pretty unclear.

So thinking about ways in which we can engage to support addressing that is probably one of the most important steps we can take for the future of water supply and it is a—it is a fear right now, right. So the discussion about climate change in the region is about this uncertainty.

Mr. KEATING. It is dynamic too because it is just not about precipitation changes. It is also flooding and other issues that result.

Ms. WOODEN. Yes, and temperature changes. So there is—the latest IPCC report had a number of instances, I think something like six or seven, indicating specific changes in Central Asia that include temperature changes that are already impacting the growing season that interacts with what is happening in the Aral Sea.

So the interaction—I don't think the conflict will happen directly but as we have all outlined, the next step to migration, for example, is one of the most important next social impacts. Where will people move to if glaciers decline? The fastest declining glaciers in the region are the lowest-lying smallest glaciers closest to population centers.

That is where we see that already happening, right, and so will people move from those locations in agricultural-dependent communities and what will that migration mean.

Mr. KEATING. We didn't touch too much on international aquifers as an issue but I think they are another concern that we should have and many of them are facing serious declines as well not just with ground water decline but with contamination.

Can you just discuss what can be done and what is being done with that, particularly in agricultural areas?

Mr. SULLIVAN. Well, if I might say something on the issue that she just touched upon, think about the following 300 to 400 million people that are reliant on the Tibetan Plateau glaciers and those glaciers might be melting. That includes three nuclear states—Pakistan, India and China.

Now, with regard to the aquifer, there is a very interesting aquifer in North Africa—the Great Nubian Sandstone aquifer, which has 150 trillion cubic meters of water in it.

Muammar al-Gaddafi—you remember him? He started something called the Great Manmade River to bring that water to the coastline to green Libya. At the same time, Egypt was tapping into the thoughts of water—this is 40,000- to 60,000-year-old water—as an alternative source of fresh water.

Once you take that stuff out, it is nonreplenishable. There is no rain. The Saudis grew wheat in the desert. The first time I flew over Saudi Arabia was in 1980 coming back from Bombay, India, writing my Ph.D. at the time.

I saw these green circles. What is this all about? Growing wheat in the desert, using the underground aquifer of nonreplenishable water. Overall, it cost about 11 times to grow the wheat than in Kansas. That was a waste.

They figure it out. They changed the program. But when people don't see what is underground they don't really think about it and when water is free just take it up, when the only cost of water to you is the diesel fuel to pump the water out of the ground.

But on the other side of the story, being from New England you can—you know how big New England is. We all do. Underneath Darfur there is a huge lake of water the size of New England, sometimes 300 meters deep.

But on top of that underground aquifer people are dying of thirst. We have to understand the situation a lot better.

Mr. KEATING. Thank you. I yield back.

Mr. MEEKS. Thank you, Mr. Chairman. And that is interesting because—what you just talked about, Mr. Sullivan.

I mean, what is striking to me is, as I said in my opening, there was a presentation on 60 Minutes just this past weekend and they were showing how deep in the ground, in California, not in places that we are talking about where folks are starving, but in California where the farmers are utilizing this ground water and pumping it up, just using the same machines as if they were digging for oil, and how the ground water is now going down, down—I think they said a foot every month now.

But the farmers claim that they need it and they were drilling more holes, et cetera. But we have got places where folks are starving and they need the water to drinks and/or, you know, you talked about what took place in Egypt.

So how do we balance the need for preserving the water and juxtapose it to those individuals who need water to drink—you know, clean water to drink, and that is clearly, you know, sanitation wise, et cetera?

How do we juxtapose we do that and working with these countries so that they can survive? You know, it is easy for—sometimes for folks to come from the United States and say we can do this because we have—you know, we are feeding our people, et cetera.

They are in a different circumstance—so that they can survive but also understand the precious resources because we all are interrelated in that regards. How do—how do we differentiate between the two? Dr. Kuehnast?

Ms. KUEHNAST. Well, I think a couple of things. One, we need good data and that needs to be shared in a transparent way from the ground level to the political elite as such.

But even more so, some of what we have to do is change daily habits and that needs very astute social marketing efforts and that is why I emphasize this, that we need, first, critical problem solving skills of the people who live there.

And second, how can we market that kind of change of behavior? Because they came out of a legacy where all things were free, where there were no incentives to save. And you have to change human behavior first to get at, you know, bridging the gap between the Darfur example so eloquently laid out here and the water underneath.

We need that bridgework—critical thinking skills, incentive programs for young people to help solve their own issues.

I loved what you said about, really, it is not only sharing but preserving water for the future—how you motivate young people to get at that problem and that helps them solve their future issues.

Mr. SULLIVAN. Well, a lot of this has to do with knowledge and data—I would agree. We really don't know what the groundwater is like in many parts of the world.

But also how many people know what their water footprint is? Two kilograms of beef takes 15,701 or thereabout liters of water and yet a similar amount of protein through vegetables would be about 1,700 liters of water.

The way we eat we have countries growing very quickly—China, for example. They are moving toward pork and beef, more water intensive ways of doing things. Energy systems—some are more water intensive than others.

The worst is biofuels, by far—irrigated biofuels—absolutely water nonsense when it comes to that, particularly in a country with low amounts of water. Getting back to Ethiopia, it is hydro dams causing all of this trouble. They have massive geothermal reserves in that country which will use less water.

They have significant solar energy potential, wind energy potential, many other potentials in that country that could be used without causing all of this trouble. At least seven gigawatts of geothermal could be rolled out in the next few years. They are on the Rift Valley.

Geothermal is a really great place when you are near hot rocks in the ground. Just north of San Francisco in your state there is a city that is run by geothermal, Calpine Geysers.

Geothermal in the world is a small percentage of what it could be. Japan could turn around to geothermal. Many countries could turn around. In this country, the biggest use of water is not irrigation.

It is thermal electric cooling—thermal electric cooling—and California is facing this as a big problem right now and many countries looking to develop their energy systems the way we did it are going to have to rethink it if they have water stress. Nuclear power plants the same thing—how do you cool them down? With water.

Mr. MEEKS. And let me ask Ms.—I think Dr. Wooden said this, that—I am just interested in the statement—you said that bilateral agreements as opposed to multilateral agreements were more effective.

You know, in my way of thinking initially was that in a region you wanted, of course, it could be interconnected between two and three different countries that you would want multilateral agreements.

I was wondering if you could give me some further clarification on why bilateral is more successful than multilateral.

Ms. WOODEN. It is interesting, in Central Asia the relationship over these two rivers has been joined. It is part of the Soviet legacy and in the post-Soviet period initially in cooperation that actually was enhanced by joint concern about the Aral Sea decline.

So actually we actually saw in the 1990s a number of high-level meetings between leaders of the countries in the region establishing an institutional framework that was rather complex that united all decisions across the region and made it difficult to separate out the individual rivers.

And so when tensions exist between two countries among the five, and six if we include Afghanistan, this makes progress in the rest of the basin difficult. And so that is why the suggestion by the international community has moved toward okay, let us break this down and work on those bilateral relationships and some of them have moved forward.

I mentioned a couple of them like the Kyrgyzstan and Kazakhstan relationship and they are not perfect. They had a little bit of a spat over water in 2010 when Kazakhstan closed the borders after the violence in Osh and Kyrgyzstan—well, there were some canals that were cut off for a little bit of time and Kazakhstan then reopened the borders.

So those bilateral relationships don't always work perfectly. But they have allowed some improvements, for example, in funding of infrastructure improvements in Kyrgyzstan from Kazakhstan, and Kazakhstan and China are experimenting with this as well.

So there was a breaking down of the complexity of the relationship to smaller bilateral agreements that also deal with some of the border delimitation issues, for example, or the enclave disputes.

And so what building cooperation also is demonstrating is that it is possible in one part of the basin. That doesn't mean that the future of moving toward multilateral agreements should be avoided. That would be preferable.

But until that is functioning—yes.

Mr. MEEKS. My last question would be I haven't heard in the testimony much and I know we in the United States have to do our part—we have to do a lot—but I didn't hear—what about the roles of—do you think in regards to the United Nations, the World Bank or other international efforts to address these freshwater conflicts?

Do they have—do they have—should they play a significant role or do we not just have confidence in them?

I haven't heard about—you know, again, it is a global context of which we are talking and so I think that we need help. But I wonder, you know, your positions on those.

Ms. WOODEN. I can mention that when it comes to the Aral Sea, many people are greatly disappointed in the limitations of the international community to assist in stopping the process of the collapse of this ecosystem.

This is—this was the focus of my dissertation—evaluating regional engagement in the Aral Sea basin—and the most successful efforts that actually surprised me were by USAID and that was because the engagements were small and spending a lot of money in a really difficult political situation is challenging.

And I think that the efforts are also sometimes incorrect. So, for example, the World Bank helped fund the construction of the Kok Aral Dam, which divides the small Aral at the north part of the sea from the larger Aral, and this was done because of the intractability of dealing with Uzbekistan's Government primarily and just in wider issues such as changing cotton production for reducing irrigation use.

And so the World Bank, when appealed to by the Kazakhstan Government, agreed to construct this dam and it worked to increase the levels of water in the small northern part of the Aral.

But it drastically sped up the decline of the rest of the lake and so this summer it no longer exists in part because of that sped up process.

So this dam is used as an example of success but it is also—you know, when we think of—we take this whole ecosystem and we break it up in parts. I just suggested doing so, right, but there are ramifications of doing that.

And so continuing to make sure we understand clearly if we engage economic growth in the textile sector what does that mean for cotton production in the wider region? We need to be aware of those ramifications.

Mr. SULLIVAN. The World Bank—excuse me. Could I—the World Bank has been involved with this but is not involved with the GERD because they have not received an environmental impact statement or an economic or social impact statement, which they require.

The GERD is actually being paid for by bonds that are being sold to the Ethiopian public at between 1.5 and 2.5 percent. There are NGOs that are really doing very good work in the small. Water.org, with Matt Damon, is doing great stuff in Africa.

Five thousand children under the age of five die every day in sub-Saharan Africa because of dirty water. You want to make friends and influence people? Clean up the water.

Mr. ROHRABACHER. Thank you very much for that admonition. I think it—we should all take that to heart. About the point you made before that—whoever it was—about the financing of a particular dam, who is financing the Rogun and the Renaissance Dams?

Who is financing that? These are the—these are the two dams where the, really, the major contention—this could end up resulting in conflict. Who is financing those?

Mr. SULLIVAN. When I say the GERD, I mean the Great Ethiopian Renaissance Dam. That is the one with the bonds. So I should have been clearer.

Mr. ROHRABACHER. So the World Bank and the other——

Mr. SULLIVAN. The World Bank is not involved. The IFC is not involved. USAID and the United States give Ethiopia about $500 million a year. There is a leverage there.

It is mostly being paid for by statement of the Ethiopian Government—by the Ethiopian Government but they are paying for it with debt. This is a country with a $50 billion GDP and this dam alone is $5 billion.

Mr. ROHRABACHER. And who is paying—who is buying the bonds?

Mr. SULLIVAN. Ethiopian expats, Ethiopians who live in the country. China is involved with the turbines. China is involved with the electricity system connections. But they don't have enough money in the country, is my guess, to do what the Egyptians did with the Suez Canal. No way.

Mr. ROHRABACHER. I know I don't have to tell you that I have— you are not giving me a certainty answer. The Ethiopian expats are buying all these bonds——

Mr. SULLIVAN. And people inside the country. They have embassy locations throughout the world. I know that sounds odd——

Mr. ROHRABACHER. It does sound odd——

Mr. SULLIVAN [continuing]. Because they couldn't find——

Mr. ROHRABACHER [continuing]. Impoverished for them to be able to——

Mr. SULLIVAN. I know. I know but——

Mr. ROHRABACHER [continuing]. Buy bonds.

Mr. SULLIVAN. But with the Suez Canal the Egyptians collected 66 billion Egyptian pounds, about $10 billion to build that.

Mr. ROHRABACHER. That seems a little more doable than the Ethiopian people who mainly live in poverty.

Mr. SULLIVAN. Okay. So now you see my lack of believing in some of the policies of the Ethiopian Government. When they are doing something like this and they are saying this——

Mr. ROHRABACHER. Yes.

Mr. SULLIVAN [continuing]. This is problematic.

Mr. ROHRABACHER. Yes. Well, let me—somebody is making money on it there.

Mr. SULLIVAN. No kidding.

Mr. ROHRABACHER. Yes, no kidding. By the way, just to mention geothermal, however, there have been drawbacks to geothermal and that is—we tried that in California. We have a large amount of geothermal and they may have solve that problem now technologically.

I would have to go back and take a look. But it destroyed the— there was a—the pipes went through a degeneration of the pipe and the material very quickly and geothermal wasn't—didn't meet its promise, let us put it that way, because there was a lot of calculation on that early on.

Now, maybe they have cured that problem without me knowing it. But I will check into it because I should know that. About the— a couple points.

We will do just a second round very quickly and with all due respect I don't think we necessarily disagree on global warming. The issue that mainly is at hand, the temperature is what the temperature is and the impacts that people measure are there.

The question is whether human activity through CO2 is causing the change in the climate. We have been through climate changes throughout the history of this planet and the only question is is the one we are at is that caused by human beings, and if it is my colleagues who believe that have—are totally justified in trying to control the behavior of human beings through government action.

If you don't think that it is human beings you end up spending enormous amounts of money and controlling people's lives basically when you should be trying to find ways of dealing with the fact that you are now in a cycle of history that will leave with less water and affect the glaciers, et cetera.

Rather than trying to control people's lives you are trying to remediate it—I guess, is the word I am looking for—the effects of that impact and whether—so we are—it sounds like we are in for some when we may face water shortages because of one part of the cycle or something that were being caused by having too many automobiles.

But whatever it is, it is there and when countries like Egypt try to deal with it and Ethiopia tried to deal with it and Central Asia, we need to put in place something, not just the answer to how we are going to lessen the suffering that may come from this but also perhaps weigh—that we put in place something that will deal with the conflicts that happen between peoples that wouldn't exist had that change in climate or the change in the status quo not happened.

We don't seem to have that. I mean, I haven't been getting that from you today as to there is something in place. Maybe we need to focus on trying to have some international mediation board or arbitration board that is signed on to any nations that have conflict and that everybody else agrees that at that point they will respect the rights and they will respect those nations that go ahead and go along with whatever the decision is.

That is one idea. Maybe there is some other ideas of how we can help countries become more efficient in the use of water and things such as that. If you have just one or two comments on that and then I will let Mr. Keating finish up with his questions.

Ms. WOODEN. On the Rogun Dam question you asked about, there are ongoing considerations—from the Tajikistani Government—about how to attract more funding and the government has in the past sought to collect funds, a forced funding collection from citizens.

So there is some controversy about how the funding has been raised and whether or not funding will be forthcoming to complete the project. So that is definitely a part of the discussion.

Regarding the possibilities of dealing with the conflicts that are produced, respectfully, I think that just like in the Aral Sea situation if we don't actually tackle the causes of the problem, if we think about temperature changes, precipitation changes, glacial decline, there are in the latest IPCC report tens of thousands of academic articles evaluated to identify pretty clearly—very clearly the pattern and the cause of anthropogenic impact, anthropogenic greenhouse gas emissions on climate change.

So we know what the causes are. There is scientific certainty and so——

Mr. ROHRABACHER. Let me just note—let me just note for the record that being a member—the vice chairman of the Science Committee that is only your opinion and the opinion of others. There is a lot of other opinions as well.

Ms. WOODEN. Well, I actually don't think it is opinion. This is scientific research, correct?

Mr. ROHRABACHER. But whether or not—whether or not it is caused by human beings or whether it is caused by a natural cycle, as I said, I don't think that it is as relevant as would be except you want to aim your solutions at what will have a good impact.

Ms. WOODEN. Right. I agree. Aiming the solutions of what will have a good impact but we have to get at the heart of the problem, and we have been asked to come here as experts in the region and plenty of experts on climatology have evaluated the problem.

So if we think about then how to deal with conflicts, just like with the Aral Sea issue the Aral Sea problem was cotton production. If we think about dealing with the problems of water supply there are ways of changing the uses of water, the ways of dealing with water pollution, for example, that makes the available water all capable of use, right.

I mean, those are really important issues to tackle that we forget when we talk about trying to increase supply in other ways, well, we have actually have to make sure that we have adequate quality of the water that exists.

So that is why we can talk about those kinds of causes and address those.

Mr. ROHRABACHER. And Dr. Sullivan, in his admonition to us about, you know, if we are really concerned about people's lives those 5,000 kids dying like that with—from bad water, sometimes—and people have this—build these grandiose projects with the dam in Ethiopia when I have been told if we just would focus on making sure that people in those villages have a way of purifying their water in sometimes very simple ways but we have to take the initiative to go out and make certain things available— that that would be much better than these grandiose dam projects.

Mr. SULLIVAN. Couldn't agree with you more on this. There is something called SODIS—S-O-D-I-S—out of Switzerland where very poor people can actually use the typical plastic bottle, filter a little bit and put it on top of a tin shack, heat the water and get rid of most of the bacteria. And then there is LifeStraw, which is pretty inexpensive—this could be handed out.

But this is kind of person to person. Another way of getting at this, particularly in remote communities is to have solar water pumps or wind water pumps, getting the cleaner water up from the aquifer.

The problem with dying from dirty water is that they are digging this out of open pit wells where the donkeys and other animals— you can take a guess at what happens and people wash and it is happening in rivers and so forth. But there are simple answers to this.

If you take a look at the $5 billion being used to bill this dam and put it toward cleaning water for under-five children in Africa you could save many American football stadiums full of African kids every single year.

Mr. ROHRABACHER. And did you have one last comment to that?

Ms. KUEHNAST. Well, I would just like to underline again that, you know, we need creative context-specific solutions. These five countries in Central Asia all have different water and energy dynamics.

One shoe does not fit all. But, you know, I am interested in the fact that Dr. Wooden said USAID's project in the Aral Sea seemed to have impact because it was locally driven, and I think so often we are attracted by the massive engineering of a dam or a road or whatever else that we lose track of the everyday efforts or technology—low-level technology that could make a difference and we need to incentivized that approach and that is my suggestion.

Mr. ROHRABACHER. Well said, because the incentive now is for some big companies to make a lot of money building these huge projects. Mr. Keating?

Mr. KEATING. Thank you, Mr. Chairman. I will just comment on some of the takeaways I get from today's hearing and you can comment as you may wish to or may not.

But the one thing that seems clear is that we don't have enough data and, to me, if we are going to work with water management and—or mismanagement and we are going to look at—deal with this issue, we seem to have a consensus among all three of you that this is a primary need and without that we are not able to move constructively.

That brings me back briefly—we won't dwell on it—to the importance of realizing as the U.N.'s Intergovernmental Panel on Climate Change has in thousands of reports that climate change—99 percent of all the scientists that poll agree that climate change is manmade and without having that basic understanding it is going to be harder to deal with that data once we get that or even accumulate that data.

So I think it is a threshold issue we have to deal with. But I want to tell you I feel optimistic after today's hearing because there is so much we can do.

This thing is not a problem that is in any stretch insurmountable or difficult if we do—if we approach it the right way. And so I came away learning that, you know, some fundamental issues about, you know, what you grow for agricultural products and where you grow it, even what we eat, are all things that have very strong impact on our water resources in this world.

I think—I mentioned just once that the government can get involved and the U.S. Government is involved in many ways. The smaller project with the Millennium Challenge in Cabo Verde, you know, we put $41 million to establish transparent water delivery and sanitation systems for that area. And so we can, you know, use some of our financial resources in that respect as well.

And I will finally just comment on, and Dr. Sullivan was stealing some of my thunder on this, but in my own district we have a military base where there was water contamination from the utilization of that and it was a major problem and it is being cleaned up, and the energy to clean that up is being generated through wind power.

So and I also have municipalities in my district that I have gone around during this break and I have seen how they are using solar and wind power in those communities to provide the energy for wastewater cleanup and for even the delivery of their own water supply.

So there is a real, you know, need, I think, because if you are doing the tradeoffs to clean up the water and to do this produces energy, takes energy. But we can do that if we commit to renewables, I think, and using it for that purpose and it is working in my district as we speak.

So I come away very hopeful on many fronts. But we do have to start with getting the data and getting the information and that is something one country can't do itself.

That is something that is going to need international cooperation with in terms of access and getting our scientists all on the same page and our engineers all on the same page and then we can approach this.

So that is, again, getting back to my introductory remarks, Mr. Chairman, why this committee has, I think, done a great service by, again, bringing up this issue that not only affects a global conflict but our survival and our assessments going forward.

So thank you very much. If any of you want to comment on any of those things you are free to.

Mr. SULLIVAN. When I think of many of the things that we talked about today sometimes I think of the little children I had seen in Egypt in my years living there. I wonder what is going to happen to them.

The water gets shorter. What happens to the women and the girls and the older people who need that water more than others? And also, how will this affect the development of terrorism and strife and the return of the Brotherhood?

Ethiopia really needs to commit publicly how quickly they are going to fill up this dam, by how much and when, and have some cooperation with Egypt and others involved to try to resolve the tensions here.

They need to do the same with Kenya for the Omo River. There has to be some precedents set to find a civilized reasonable way of solving these issues without conflict because it goes right back to the little kids in the street in Cairo.

Ms. WOODEN. I would like to add on the data sharing issue or the data collection issue that in part it is to understand some of these processes are happening—to be able to understand them well—for us here to understand them—for decision makers in the region to know what is happening in the future.

But also data sharing is an important part of intergovernmental relations, right. I mean, it is one of the biggest problems that we have between Uzbekistan and Tajikistan and Kyrgyzstan that is that transparency about these processes.

Ability to predict mutually understood change and how to adapt to it as it is happening is very important. I mean, the understanding of how we can monitor from satellite imagery, monitor ground water changes in California has been really important for our ability to withstand—in a significant drought withstand severe political ramifications, right. That is what we are talking about happening in Central Asia.

A big part of this is just being able to make this whole process transparent. So I think funding of research is one of the most crucial steps we can take to tackle this and to generate cooperation.

We have already talked about cooperation as happening more commonly, right. So there is much to build on here in the relationships between communities on the border and between governments and we can really begin by just listening to leaders but community leaders at the local level for what they need.

Most people in the region are concerned about water as the primary issue. We both have found this in our surveys of the region—that water pollution and water supply are primary concerns.

So people know what the issues are, are worried about them and want to work together to tackle them so mainly through improved understanding of what is happening.

Ms. KUEHNAST. I would like to say that, indeed, transparency of information is critical. But what you need is the investment of the young people in this five-country region with a sense of hope, with a sense that they can apply good knowledge with excellent business and technological acumen and help solve their problems, help strengthen and build capacity at the local level and I think you will see more wind farms, more solar energy, more direct person-to-person and technology advancements that are really responsive to the issues and in doing so you will prevent conflict because you give people the sense that they can take care of this themselves and that they are empowered to do so.

Thank you.

Mr. ROHRABACHER. Well, thank you all, and just a note that, again, I want to say very clearly that the United States is watching Ethiopia's activities and decisions on this dam project. We are watching very closely.

We are—we would expect, if Ethiopia is acting responsibly, that it commit itself to publicly to a policy that will ensure that they are not doing a project that will benefit them at the expense of the people of Egypt but instead will try to work and hopefully work with the Egyptians to find water solutions that do not harm large numbers of other people who happen to live across the border or downstream from you, and this—the Ethiopian Government better understand that or there will be major retaliation from this Congress on Ethiopia for that type of hostile act toward the people of Egypt.

In terms of Central Asia, I would hope that Uzbekistan and Tajikistan as well as these Central Asian countries will be able to—because they all are dependent on water, I mean, especially with Tajikistan and Uzbekistan should understand this problem and hopefully they will work together and that there will be a Central Asian cooperative spirit that will put people together for—to come up with an understanding on this, and maybe we can play a positive role in all of this by thinking about establishing, as your organization is aimed at, trying to find out how we can as a people serve as conflict deterrents and how we can become active in a process of deterring conflict by being arbiters or being people who are—we at least could come in and get the parties together and find ways of reaching agreements between people who have disagreements that might lead to conflict and water is, of course, one of those major issues.

I do believe that we should continue funding. One of the reasons why I am upset with the focus on human activity is that it—we have spent billions of dollars trying to determine what will be the impact of global warming and with the idea of justifying the expenditures on that research.

I think that we should instead have research into finding ways that are going to make people's lives better that will actually be able to offer some sort of impact on those people's lives there, whether it is the water pollution devices that Dr. Sullivan talked about or other types of technologies that will permit people in Africa to get—to cheaply get to cleaner water, which seems to be better than building huge dam projects, et cetera.

So but, again, whether or not it is caused by human beings or whether it is caused by—one last note. That 99 percent figure has been figure has been disproven over and over and over again.

Ninety-9 percent of the scientists do not agree that mankind is causing this change in the climate. It is a majority, however do agree with you and disagree with but not 99 percent.

And with that said, I want to thank the witnesses. We have had a very good discussion and I have really always felt that there are two major important things for people—to be able to have a planet where ordinary people are going to live decent lives.

We have got to have energy and hopefully clean energy and we have got to have water, and with those two things I think human beings and human ingenuity will be able to overcome a lot of other things and develop the agriculture, et cetera, that we need. But without those two fundamental things in play, ordinary people won't live well.

So I think the United States should be committed to clean energy and water for the world.

Thank you very much. This hearing is adjourned.

[Whereupon, at 3:40 p.m., the subcommittee was adjourned.]

APPENDIX

MATERIAL SUBMITTED FOR THE RECORD

SUBCOMMITTEE HEARING NOTICE
COMMITTEE ON FOREIGN AFFAIRS
U.S. HOUSE OF REPRESENTATIVES
WASHINGTON, DC 20515-6128

Subcommittee on Europe, Eurasia, and Emerging Threats
Dana Rohrabacher (R-CA), Chairman

November 17, 2014

TO: MEMBERS OF THE COMMITTEE ON FOREIGN AFFAIRS

You are respectfully requested to attend an OPEN hearing of the Committee on Foreign Affairs, to be held by the Subcommittee on Europe, Eurasia, and Emerging Threats in Room 2255 of the Rayburn House Office Building (and available live on the Committee website at http://www.ForeignAffairs.house.gov):

DATE: Tuesday, November 18, 2014

TIME: 2:00 p.m.

SUBJECT: Water Sharing Conflicts and the Threat to International Peace

WITNESSES: Paul Sullivan, Ph.D.
                   Professor of Economics
                   National Defense University

                   Amanda Wooden, Ph.D.
                   Associate Professor of Environmental Studies
                   Bucknell University

                   Kathleen Kuehnast, Ph.D.
                   Director
                   Center for Gender & Peacebuilding
                   United States Institute of Peace

By Direction of the Chairman

COMMITTEE ON FOREIGN AFFAIRS

MINUTES OF SUBCOMMITTEE ON _____ *Europe, Eurasia, and Emerging Threats* _____ HEARING

Day___ *Tuesday* ___Date_____ *11/18/14* _____Room___ *Rayburn 2255* ___

Starting Time ___ *2:04PM* ___Ending Time ___ *3:39PM* ___

Recesses |___| (___to ___)(___to ___)(___to ___)(___to ___)(___to ___)(___to ___)

Presiding Member(s)

Rep. Rohrabacher

Check all of the following that apply:

Open Session ☑
Executive (closed) Session ☐
Televised ☐

Electronically Recorded (taped) ☑
Stenographic Record ☑

TITLE OF HEARING:

Water Sharing Conflicts & The Threat to International Peace

SUBCOMMITTEE MEMBERS PRESENT:

Rep. Meeks
Rep. Keating

NON-SUBCOMMITTEE MEMBERS PRESENT: *(Mark with an * if they are not members of full committee.)*

HEARING WITNESSES: Same as meeting notice attached? Yes ☑ No ☐
(If "no", please list below and include title, agency, department, or organization.)

STATEMENTS FOR THE RECORD: *(List any statements submitted for the record.)*

TIME SCHEDULED TO RECONVENE _____
or
TIME ADJOURNED ___ *3:39PM* ___

Subcommittee Staff Director